Studies in Computational Intelligence

Volume 928

Series Editor

Janusz Kacprzyk, Polish Academy of Sciences, Warsaw, Poland

The series "Studies in Computational Intelligence" (SCI) publishes new developments and advances in the various areas of computational intelligence—quickly and with a high quality. The intent is to cover the theory, applications, and design methods of computational intelligence, as embedded in the fields of engineering, computer science, physics and life sciences, as well as the methodologies behind them. The series contains monographs, lecture notes and edited volumes in computational intelligence spanning the areas of neural networks, connectionist systems, genetic algorithms, evolutionary computation, artificial intelligence, cellular automata, self-organizing systems, soft computing, fuzzy systems, and hybrid intelligent systems. Of particular value to both the contributors and the readership are the short publication timeframe and the world-wide distribution, which enable both wide and rapid dissemination of research output.

Indexed by SCOPUS, DBLP, WTI Frankfurt eG, zbMATH, SCImago.

All books published in the series are submitted for consideration in Web of Science.

More information about this series at http://www.springer.com/series/7092

Alexiei Dingli · Foaad Haddod ·
Christina Klüver
Editors

Artificial Intelligence in Industry 4.0

A Collection of Innovative Research Case-studies that are Reworking the Way We Look at Industry 4.0 Thanks to Artificial Intelligence

Springer

Editors
Alexiei Dingli
Faculty of ICT
Department of Artificial Intelligence
University of Malta
Msida, Malta

Foaad Haddod
Faculty of ICT
Department of Artificial Intelligence
University of Malta
Msida, Malta

Christina Klüver
Universität Duisburg-Essen
Essen, Germany

ISSN 1860-949X ISSN 1860-9503 (electronic)
Studies in Computational Intelligence
ISBN 978-3-030-61044-9 ISBN 978-3-030-61045-6 (eBook)
https://doi.org/10.1007/978-3-030-61045-6

This Springer imprint is published by the registered company Springer Nature Switzerland AG
The registered company address is: Gewerbestrasse 11, 6330 Cham, Switzerland

This book is dedicated to all those who are prepared to tread new paths. May their work usher the rise of technologies which create a whole new world. One where poverty and inequalities are wiped away forever.

Alexiei Dingli
Foaad Haddod
Christina Klüver

Foreword

AI is the discipline that studies the design, the development, and the construction of systems that simulate human skills, reasoning, and behaviour. If the term AI has a date of birth, it could be the Dartmouth Summer Research Conference in 1956. From this event, AI evolved from mere syntactic manipulation towards the development of semantic knowledge, especially over the last three decades.

The "acceleration" was made possible thanks to the development of the nanotechnologies enabling the enormous computer capability made available—think of the PC evolution beginning in the 80s and web-based solutions from the late 90s. We are still far from the "strong AI" level. The current stage of AI is what is commonly called "Augmented Intelligence" and represents for Industry an enormous opportunity to enhance manufacturing competitiveness.

Industries' interest in AI lies in its core technological function to be used, in principle, to implement every formalisation of human and rational thinking and acting, including learning, planning, and problem solving. With this potential, researchers and technical savants see AI in the group of "disruptive" technologies of exceptional potential with benefit to Industrial (and other) systems.

In Industry 4.0 processes, AI shows great potential to add value in, and provide competitive advantage to, manufacturing. On the basis of data and through skilful connections, AI can reduce complexity and detect events or patterns, even without explicit programming or classic automation and control engineering. Moreover, the steady adoption of Industry 4.0 and Industrial digitalisation is automating the production of a large quantity of valuable data that is exactly the data needed by AI to generate value.

So manufacturers are looking at AI technologies to increase the efficiency and the effectiveness of industrial processes, as measure by traditional factors like cost, speed, precision, and problem solving. They expect they can achieve higher levels of autonomy through cognitive capacities supplied by AI. And they can reach this autonomy with AI beyond what human beings are capable of.

The traditional focus on product-oriented production and manufacturing in Industry 4.0 will be soon supplanted by concepts of a more solution and customer-oriented nature. This will transform the rigidly defined chains of productions into flexible and highly dynamic manufacturing systems. These will encourage autonomously organised production systems optimised with the help of AI to meet customised Customer requirements.

It is not a stretch to say that AI will transform the relationship between people and technology, whilst boosting creativity and capability. Whilst quantifying the impact of these changes on economic and social systems is very complex, due to the number of variables, McKinsey Global studies estimate that by 2030, AI could increase global economic activity by about 13B$, a value similar to the Steam Power revolution delivered to the nineteenth century or Robots and Computers provided one century later.

For Industry, AI is not an option. It's the new way to achieve Excellence.

August 2020

Alfio Giovanni Pio Di Batolo
BEM&T Supply Chain Director
ST Microelectronics
Catania, Italy

Preface

The use of Artificial Intelligence (AI) in Industry 4.0 has been baking for quite a whilst, but technology is now ripe enough to turn it into reality. It involves the infusion of data and smart automation within the manufacturing process together with human input. To do so, it makes use of various technologies such as Internet of Things, Cloud Computing, Blockchain and others. Let us try for a second to understand why the use of AI is important.

If you are the owner of a manufacturing plant, then you're probably familiar with having various machines working in tandem and producing thousands of devices every day. Monitoring the different processes which go on simultaneously is already a huge headache. Most probably, you are at the mercy of the various operators in order to ensure that the level of production is at least maintained. However, even though you are taking the best possible measures to mitigate any risks and ensure that everything runs smoothly, things still go wrong. When they do, havoc breaks loose until the main issues are sorted out. Things get further complicated when problems arise during a weekend or a public holiday since the availability of skilled professionals is very limited and costly. When production is halted, it obviously results in financial losses which might also impact production deadlines. Large production processes would normally depend on warehouses of raw materials, parts or other items. All of these cost a lot of money and involve huge risks if they are mismanaged. The handling of these assets alone is already a huge headache for management. And we haven't even started considering how to improve the plant; maybe through optimisation or changing the basic configuration. But sometimes, when dealing with legacy systems, any change can be catastrophic and as such, management might be slow to respond towards the changing needs of the market. Sometimes, it is simply too late and the lack of decisive action leads the plant towards certain doom.

Do these scenarios sound too familiar? Well, the good thing is that most of them can be handled quite easily with today's technologies. To understand what we mean, let us give you some examples.

Micro-Monitoring at a Distance

Consider managing a plant with hundreds of machines, working like clockwork. To complicate matters, let's imagine that there is more than one plant distributed around the world and located in different time-zones. How can you manage all of that? The answer is that one can't micromanage it! You'll have people to do that, especially when the numbers start growing. But we all know that people are not infallible. The solution to that is to instal an Intelligent Digital Twin (IDT) system capable of monitoring all of your machines with 24/7 constant coverage, irrespective of where they are physically located. You can forget about it and let the system take care of the issues which arise, whilst only alerting you when your attention is really needed. Furthermore, it will give you the peace-of-mind which you need since you can zoom into the inner workings of any machine when you want and in real time. With the advent of tools like Virtual Realty (VR) headsets or Augmented Reality (AR) glasses, you can be either virtually transported to the manufacturing plant using VR or the plant is virtually transported to your office using AR. A mixed reality experience can be enjoyed anywhere in the world and allows you to see in real time each and every part of your facility.

Optimising your Plant

Plant optimisation and improvement is a big game changer. However, it is frequently ignored because as the saying goes "if it ain't broke, don't fix it". But how true is that? Did anyone ever try calculating the opportunity loss of doing things in a different way? Most probably, it never happens because normally, you would have to rely on experts to drill down to that information. Unless experts have time to spare (which they normally don't) or if they are not prompted by management, they would never do it. But management seldom looks at these things because they are busy with the day-to-day running of the plant, so things keep on deteriorating slowing until the day when the system suddenly fails or when that competitive edge starts disappearing due to inefficiencies in the system.

With AI, this scenario changes because the system actually performs these calculations without needing to be prompted by anyone. It will constantly analyse the supply chain and quickly simulate the ideal and most efficient scenario for any particular job. It does so by looking at the current production, predicts potential down time, slot in new tasks in the pipeline and make suggestions based upon deadlines such as delivery targets and material available. All at the click-of-a-button. It is the ultimate aim of the AI to ensure that the production plant is optimised at all time and it is constantly readjusting those optimisation parameters in real time based upon changes which occur on the factory floor.

Predicting the Future

Predicting the future is not something easy but it is a very desirable feature especially when one considers that issues such as machine downtime is one of the largest contributors to loss of production. In fact, 80% of companies are unable to calculate their true downtime costs. On average, companies record 4 hours of downtime daily! Many have tried to tackle this feat but very few predictions can really survive the test of time. However, when dealing with a closed system, past events can give a very good indication of what will happen in the future. Machines in a manufacturing plant can be considered as a closed system which have been running for months and in some cases, for years. Being mechanical in nature, they constantly log heaps of precious diagnostic data which is probably lying in some database just in case the engineers need it. In reality, sifting through all that data is rather painful, precisely because of its volume, so people seldom refer to it.

However, AI systems excel at processing huge volumes of data and analysing it. Specifically, they find it rather easy to locate patterns and these patterns can give a good indication of what is going to happen. Hence, an AI system can easily make predictions on when the machine is going to fail and for what reason. This will allow the plant to organise prescriptive maintenance which shifts the management's strict dependence on planned maintenance, to being able to take real-time action based upon actual events. Furthermore, the system will not only predict a potential failure but it will also help management identify the root cause of the issue. Thus, some of the failures might be eliminated from the production cycle once and for all, whilst others will be handled in a timely manner. In so doing, unplanned downtime can be heavily reduced or even eliminate in some cases, thus maximising profitability and equipment reliability.

A Factory Made of Lego

In an ideal world, one would gather the machines and move them around until the right configuration is found. This would minimise costs and optimise the available space. Furthermore, when the plant starts growing and new machines are added, different configurations might be beneficial. In reality this does not happen because it takes a lot of time and effort to do so, probably so much that it outweighs the benefits.

However, with an IDT system, this can be achieved rather easily. An IDT system already has a virtual replica of all the machines and their status is updated in real time. Changing the configuration of a system thus becomes child's play since moving virtual machines and changing configurations is similar like playing with Lego. Machines can be added or removed and their output can be easily simulated using the historic information which the system already possesses. Thus, different trials can be setup, either by the managers or by the AI system controlling the plant, in order to find out the best configuration possible. Once done, the value of such a configuration is assessed and if it outweighs the costs, it can be deployed in the physical system on the factory floor.

Asset-Management Made Easy

Managing all the assets in an organisation is a big headache especially in plants having hundreds, if not thousands, of machines working simultaneously. Because of this, predictions become even more valuable especially when managing the purchasing of spare parts in order to replace defective components or when planning for forthcoming machine reconfigurations. When doing so, Artificial Intelligence systems can be used to scrape through the online stores, estimate the cost of parts (taking into consideration delivery dates, etc.) and predict the total cost of ownership.

All the components; both the purchased ones and those produced through the manufacturing process can be tracked through their unique identifier. Some of them will be tagged using technologies such as Radio Frequency IDs (RFIDs) which seamlessly provide information to the central system with regard to the physical movement of components. Thus, with the click-of-a-button, the person managing the system will have an up-to-date overview of all the physical stock available in the plant at any one point in time. Long gone are the days of never-ending stocktaking. The system will provide additional information such as products on order or those in transit. Furthermore, if the tracking module is built upon a blockchain system, one can ensure full traceability for any product; starting from the raw materials which make up the product, up to the end of the product's lifecycle in the recycling stage.

Better Decision-Making

IDT systems will be there to assist management in the provision of timely information, in the sifting through the sea of data and in creation of summarised viewpoints over the operations of the various plants. Without doubt, they will become the personal assistants of management capable of not only giving accurate snapshots but also highlighting areas of concern in real-time.

Furthermore, the time will come when the management of the organisation will be entrusted in the hands of the AI system thus automating the entire decision-making process. Decisions will be taken at the speed of light and corrective actions will be dispatched as soon as something happens. Managers will be relieved from the day-to-day micro-management, they can look at the plant from the macro-perspective and spend more time on planning future improvements in line with the direction imparted by the board of directors.

Conclusion

The future looks bright and exciting for the industries of tomorrow. But the future starts today since most of the technologies mentioned above are already available. In this book, we will showcase further advancements in the field of AI in order to push forward the digital transformation that will change our industries forever.

Msida, Malta — Alexiei Dingli
Msida, Malta — Foaad Haddod
Essen, Germany — Christina Klüver
August 2020

Contents

About the Editors

Alexiei Dingli is a Professor of Artificial Intelligence (AI). He has been conducting research and working in the field of AI for the past two decades. His work was rated World Class by international experts and won various prizes including the first prize by the European Space Agency, the first Prize in the Malta Innovation Awards, the World Intellectual Property Organisation (WIPO) award for Creativity and the first prize of the Energy Globe award by the UN amongst others. He has published several peer-reviewed papers and books in the field. He is also involved in various AI projects with local and multinational organisations on the implementation of Industry 4.0 system and Intelligent Digital Twins architectures. The Government of Malta appointed him on the Malta.AI task-force aimed at making Malta one of the top AI countries in the world.

Foaad Haddod is a Ph.D. student and researcher at the Department of Artificial Intelligence, Faculty of ICT, The University of Malta. He has been working on an industrial collaboration between STMicroelectronics (Malta) and the University of Malta in order to develop smart tools and systems which make use of Machine Learning, Big Data and Data Visualisation. His current work focuses on the development of intelligent manufacturing systems using smart technologies such as Digital Twins and Industry 4.0. He has published several journal articles and presented at professional conferences, exhibitions and other events held around Europe. He also designed and developed a novel Virtual Reality application to help Teachers and LSAs better understand autism.

Christina Klüver is a private Lecturer in Soft Computing at the University of Duisburg-Essen, and CEO of the own consulting company for Artificial Intelligence and Artificial Life. She has published several peer-reviewed papers and books in these fields. Together with Jürgen Klüver she has developed new algorithms in these fields as the Self-Enforcing Network (SEN), Regulatory-Algorithm (RGA) and Algorithm for Neighbourhood Generating (ANG). The main research is the Computer-based Analysis of Social and Cognitive Complexity.

Acronyms

ABS	Agent-Based Simulation
ADAMAAS-I	Adaptive and Mobile Action Assistance for Industry
AI	Artificial Intelligence
AIT	Artificial Intelligence Technologies
AmI	Ambient Intelligence
AWOSE	Agile Worth-oriented Systems Engineering
CA	Conveyor Agent
CASPA	Correct Action Selection Probability Analysis
CMP	Context Modelling Profile
CNN	Convolutional Neural Networks
CPMC	Cyber-Physical Manufacturing Cloud
CPPS	Cyber-Physical Production System
CPS	Cyber-Physical Systems
CRM	Customer Relationship Management
CSCW	Computer Supported Cooperative Work
DDPG	Deep Deterministic Policy Gradient
DES	Discrete-Event Simulation
DT	Digital Technologies
DT	Digital Twin
ERP	Enterprise Resource Planning
FPGA	Field Programmable Gate Arrays
GPU	Graphical Processing Unit
HHI	Human-Human Interaction
HMI	Human-Machine Interaction
HRM	Human Resource Management
ICT	Information Communication Technology
IDSS	Intelligent Decision Support Systems
IoS	Internet of Service
IoT	Internet of Things
LSTM	Long Short-Term Memory

M2M	Machine to Machine Communication
MA	Machining Agent
MARL	Multi-Agent Reinforcement Learning
MAS	Multi-Agent System
MDA	Model Driven Architecture
MDD	Model Driven Development
MDP	Markov Decision Process
MHI	Machine-Human Interaction
ML	Machine Learning
MOOC	Massive Online Open Course
MP	Model of Procedure
MSI	Multi-Source Identity
PA	Product Agent
PPO	Policy Optimization Algorithm
SD	System Dynamics
SEN	Self-Enforcing Networks
SME	Small and Medium-sized Enterprises
SMOTE	Synthetic Minority Over-sampling Technique
SSI	Self-Sovereign Identity
TA	Task Agent
VA	Virtual Assistants
VR	Virtual Reality
XAI	Explainable Artificial Intelligence

List of Figures

List of Tables

Part I
Managing the Industry 4.0 Revolution

We must deal quickly with the fusion of the online world and the world of industrial production. In Germany, we call it Industrie 4.0.

Angela Merkel
Chancellor of Germany

Chapter 1
Implementing Industry 4.0—The Need for a Holistic Approach

Christian Hoyer, Indra Gunawan, and Carmen Haule Reaiche

Abstract Recent studies have shown that corporations around the world are aware of the topic industry 4.0 and the possible implications associated with it, such as more flexible, resilient, and productive manufacturing sites. While the outlook seems positive, the findings of several studies have also documented that most of the companies that are aware of industry 4.0 technologies and concepts still have not started to implement them. At the same time, corporations that are already engaged with the transition faced a considerable number of challenges. These challenges have been the subject of several investigations since the introduction of Industry 4.0 and illustrate why corporations seem to hesitate with initiating the transitional process. What is more, these findings also show that while some factors slow down the adaptation of new technologies and concepts related to Industry 4.0, other factors seem to have positive effects and even make a potential implementation more likely. However, to the frustration of practitioners, scientists, and governments, the current landscape of Industry 4.0 implementation research resembles a puzzle that has not been put together yet. As a result, the current status of the implementation of Industry 4.0 is difficult to comprehend at best as the referring knowledge and expertise are widely dispersed throughout a number of different publications. Furthermore, the potential dynamic between implementation factors cannot be examined if elements are treated as isolated and individual phenomena. In order to shed more light on this shortfall, this study set out to explore the factors that have an impact on the implementation of industry 4.0 and to build a solid foundation for future research that is mainly focused on a more coherent and complete understanding of the transition towards Industry 4.0 by acknowledging the dynamic between the already identified factors.

C. Hoyer · I. Gunawan (✉) · C. H. Reaiche
Business School, University of Adelaide, 9/10 Pulteney St, Adelaide, SA 5005, Australia
e-mail: indra.gunawan@adelaide.edu.au

C. Hoyer
e-mail: christian.hoyer@adelaide.edu.au

C. H. Reaiche
e-mail: carmen.reaiche@adelaide.edu.au

A. Dingli et al. (eds.), *Artificial Intelligence in Industry 4.0*,
Studies in Computational Intelligence 928,
https://doi.org/10.1007/978-3-030-61045-6_1

1.1 Introduction

Industry 4.0, also referred to as the Fourth Industrial Revolution, gained international attention after its introduction in the year 2011. This attention is not only coming from the scientific world, which is mostly engaged with technological aspects of Industry 4.0, but also from corporations and governments. While corporations are interested in how Industry 4.0 can make them more efficient and responsive, governmental institutions, on the other hand, are mainly focused on supporting their local business infrastructures in order to save and create jobs [11, 20, 30]. In this regard, some findings suggest that the technical focus of scientists results from the need of corporations to understand the potential and usability of new technological innovations before they consider integrating them [13]. However, studies also have shown that identifying the potential of certain technologies is not the only factor corporations consider when it comes to the implementation of Industry 4.0 [17, 18, 28]. In this context, a wide variety of potential factors that have an impact on the implementation of Industry 4.0 have been identified within the Industry 4.0 literature. These factors not only potentially make the implementation of Industry 4.0 more or less likely but also more or less successful. A comprehensive list of the factors as well as the knowledge illustrating why these factors are relevant in the context of implementing Industry 4.0, therefore, can help corporations, scientists, and governments to deal with the transition towards Industry 4.0 more efficiently. Yet, an investigation which compiles and validates the factors that have already been identified has not been carried out. Therefore, this paper will discuss how future investigations can close the resulting gap and shift the focus of Industry 4.0 more towards its implementation by creating a foundation for future research.

After an initial review of the literature, different categories of Industry 4.0 research emerged. One the one hand, Industry 4.0 readiness studies have presented different areas companies should focus on before implementing Industry 4.0. These studies are particularly focused on the measures that can be taken by corporations such as investing in a more open IT-infrastructure or in the promotion of employees. On the other hand, studies that assessed the opportunities and challenges of an Industry 4.0 implementation also presented external perspectives, such as the role of the government or the standardization of communication. The third category of studies focused on the implementation of Industry 4.0 technologies in certain environments, such as a production line of an automobile manufacturer. These studies offer a better understanding of the benefits corporations are expecting from implementing Industry 4.0 as well as which obstacles they faced during the transition. However, little attention has been paid to combine these findings in order to gain a comprehensive understanding of the factors that have an impact on the Implementation of Industry 4.0. The following section, therefore, provides an overview of relevant research areas before the first findings of the resulting systematic literature will be presented. Based on these findings, further questions that need to be addressed in future research will be identified and discussed.

1.2 Related Works

The status quo of the implementation of industry 4.0 can be derived from a number of different studies. This section will, therefore, present the most relevant categories of Industry 4.0 implementation research in order to illustrate the variety of potential sources and to facilitate future research undertakings. This section will illustrate that Industry 4.0 can be observed from various angles. Identifying these angles as well as bringing them together is crucial for the understanding of the big picture behind Industry 4.0 and its implementation.

1.2.1 Implementation Studies

Large-scale Industry 4.0 implementation studies offer a clear viewing angle on the topic and show the number of corporations that already have started with the transition towards Industry 4.0. In this context, several investigations suggest that the majority of corporations start to implement Industry 4.0 by initiating isolated projects [14, 27]. Corporations, therefore, first want to understand the underlying dynamics of an Industry 4.0 implementation before they start to choose a more comprehensive approach that encompasses the whole infrastructure. A plausible decision, as several studies came to the conclusion that corporations that have begun to implement Industry 4.0 struggle to achieve higher degrees. Other studies, in contrast, have been mainly focused on the challenges and opportunities of an Industry 4.0 implementation. Those studies offer possible explanations for what makes reaching higher degrees of Industry 4.0 difficult as well as why do companies hesitate to implement Industry 4.0 in the first place. For instance, the results of the empirical investigation of Lin et al. [12] challenged the view that the size of the company matters when it comes to the transition towards Industry 4.0. Nevertheless, previous investigations have shown the link between the size of a company and Industry 4.0 adaptation, illustrating the need for further investigation about the relevance of certain factors that already have been identified [18, 26]. The third category of studies that help to understand the dynamics of adapting Industry 4.0 is the category Industry 4.0 readiness indices. Investigations belonging to this stream of research typically identify areas companies need to improve in before they start to implement Industry 4.0. For example, Schumacher et al. [7] developed a maturity model that helps corporations to assess their own infrastructures to identify potential issues that need to be resolved before engaging with the transition towards Industry 4.0.

1.2.2 *Industry 4.0 Technologies*

While the previously mentioned categories of studies are directly tied to the implementation of Industry 4.0 and, therefore, provide a good overview of potential implementation factors, they are not representing the dominant part of research in that field. Instead, the majority of Industry 4.0 studies are rather focused on providing solutions for specific technical issues that might, or might not, be relevant for companies engaging in the transition towards Industry 4.0. In this sense, other factors such as the sector corporations are operating in or the compatibility with their infrastructure play a more dominant role. However, this does not mean that these studies can also be considered irrelevant for the research revolving around the implementation of Industry 4.0. They still present findings that allow drawing conclusions about what companies need to consider once they embark on implementing Industry 4.0 into their systems and processes. In this context, the study of Frank et al. [8] shows how findings on Industry 4.0 related technologies can be translated into concrete recommendations of actions on a more general basis that provides orientation for companies regardless of the technologies they intend to integrate into their own infrastructure. Consequently, this in its nature inductive approach can lead to further findings for a larger group of companies. What is more, it also allows taking studies into consideration that are mainly focused on technologies that regarded as Industry 4.0 technologies such as Big-Data and machine learning [10, 23, 31].

1.3 The Ancestors of Industry 4.0

The last categories of studies that can be considered relevant for the transition towards Industry 4.0 are studies that examine past industrial revolutions and studies that focus on the implementation of lean manufacturing. While both categories are not directly addressing Industry 4.0 by itself, they both allow making predictions about the overall development of Industry 4.0. For example, Maghazei and Netland [13] have shown parallels between the third and fourth industrial revolution by comparing the general approach taken by scientists. In both cases, the majority of published papers first examined the usefulness of potential technologies before the attention started to shift in favour of more implementation focused examinations.

On the other hand, building on the fact that Industry 4.0 is not mainly about technologies but on adapting and developing a new philosophy that helps to organize work more efficiently, scientist started to examine the connection between Lean Manufacturing and Industry 4.0 [2, 19, 20, 24]. Both approaches have far-reaching consequences for those who are trying to implement them into their companies [21]. Therefore, it is no surprise that scientists already have started to examine the transitional process of Lean Manufacturing in order to derive lessons that can help to improve the adaptation of Industry 4.0 technologies and, more importantly, principles. However, several investigations suggest that this relationship is not one-directional as Lean Manufacturing also benefits from the advancements made in the field of Industry 4.0 [28].

1.4 Research Methods

As pointed out in the introduction to this paper, no single study exists that combined the recent findings which have been presented in the previous section. However, the combination of those findings is necessary in order to identify potential gaps that result from contradictory findings. Furthermore, an extensive list of factors that are important in the context of the implementation of Industry 4.0 has not been compiled so far. This list of factors could not only be used as a foundation for future research, but also serve as a map which gives orientation to all parties involved in the transition towards Industry 4.0 including corporations, governmental institutions, and universities. The tools that will be used to stake out that map will be presented in the following sections.

1.4.1 Systematic Literature Review

Systematic literature reviews help to identify and transparently select resources for a given topic. In order to maintain a predefined level of transparency, they follow a specific procedure that will be briefly described along with its four main phases [15, 29]. Firstly, appropriate search terms and research databases, as well as database settings, need to be selected based on a preliminary review of the literature. Secondly, filters must be defined to select papers that are relevant to the context of the investigation. Thirdly, the selected search terms and databases now will return scientific publications based on the predefined selection criteria. The filters, on the other hand, will then be used to remove duplicates and papers that are classified as not relevant after screening the titles and abstracts. Finally, the remaining articles will be thoroughly analysed in order to identify potential Industry 4.0 implementation factors. Papers that do not mention implementation factors or did not explain why those factors may be relevant will not be considered for the final analysis.

1.4.2 Qualitative Approach

Building on the results of the systematic literature review, qualitative and quantitative investigations will be conducted to validate the findings from the systematic literature review. In this context, interviewing participants who have been, or still are, involved in the transition towards Industry 4.0 provides insights into their motivation to get engaged with the topic. However, further selection criteria such as the current job position of the participant and the sector his/her organization is operating in will be applied in order to make sure that the research reflects the variety of companies, universities and governmental institutions involved in the implementation process. The interviews will then be used to identify the perceived benefits and the obstacles the participating individuals together with their organizations have faced as well as how their perception has changed over time. The resulting information will be translated into factors and compared with the results from the systematic literature review and the survey.

1.4.3 Quantitative Approach

The goal of the survey, on the other hand, is to find out how important the factors are that have been identified within the existing literature. However, since the implementation of Industry 4.0 is a long process, different implementation stages along the overall timeline will be introduced to participants. This allows them not just to state how important they perceive a given factor, but also when they think this factor becomes the most relevant. In other words, participants will also be asked to state when certain factors become particularly important during the transitional process. This is important as certain factors might be more critical at the beginning of the implementation phase, while other factors might become relevant towards the end of the transition.

1.4.4 Triangulation

By triangulating the results of the systematic literature review, the interviews, and the survey, a framework can be designed which is tailored around the question which factors are important in the context of implementing Industry 4.0. As previously mentioned, a multidimensional approach will be applied to make sure that the dynamic nature of the implementation process is sufficiently considered. In this respect, several studies have suggested that companies are at different stages of implementing Industry 4.0, which is why it is crucial to design a framework that takes these conditions into account [6]. However, while the needs of corporations build the core of the framework, it will also be addressed to other key players in the field. For instance, the framework can help scientists to get a better understanding of the research areas and factors that deserve more attention. Governmental institutions, on the other hand, can address the problems of their local business environment more accurately and as a result, contribute to more efficient funding measures.

1.5 Discussion

The systematic literature review identified fourteen factors (Table 1.1) that may be important in the context of the implementation of Industry 4.0. To illustrate the nature of those factors, three main categories have been designed from the perspective of corporations: External factors, internal factors, and company characteristics. However, the proposed categories are only a starting point as other categories might emerge throughout the course of the investigation.

1.5.1 External Factors

External factors are factors that are out of the reach of corporations' control. In other words, external factors cannot be influenced by the actions of single corporations. For instance, even though studies suggest that funding options matter in the context of implementing Industry 4.0, companies are not able to directly increase the number of available options. Hence, external factors particularly deserve the attention of governmental institutions and initiatives that want to support companies with the transitional process.

1.5.2 Internal Factors

In contrast, internal factors are factors that can be directly influenced by the actions of corporations. For example, as the review of the literature shows, the promotion of Industry 4.0 knowledge and skills might be relevant regarding an Industry 4.0 adaptation. However, companies themselves have to decide to which degree they promote these skills. Even though internal factors mainly fall under the responsibility of the organizations themselves, Industry 4.0 initiatives and governmental institutions can still have either a positive or negative impact on organizations' ability to address

Table 1.1 Industry 4.0—implementation factors

External factors	Internal factors	Company characteristics
Political support	Perceived benefits	Industry sector
IT-standardization	Strategic consideration	Company size
Institutional cooperation	IT-maturity	
Available funding options	Skills development	
Available knowledge	Lean manufacturing experience	
Pressure to adapt	Occupational health and safety	

these factors. For example, while the factor IT-Maturity mainly refers to the current state of the IT-Infrastructure in terms of flexibility and skill level, governments can initiate programs that facilitate and encourage the migration of foreign IT-experts. As a result, companies would be given more options to find specialized IT-workers [25].

1.5.3 Company Characteristics

Company characteristics, on the other hand, is a category that lists all the factors that are considered as neutral such as the size of a company or the sector it is operating in. While these factors cannot be influenced by external nor internal actions as they have a descriptive function, they still hold valuable information when it comes to the implementation of Industry 4.0. This became particularly clear after reviewing the relevant literature, which has shown that different industry sectors are more likely to implement Industry 4.0 than others. A fact that can help governmental institutions to develop Industry 4.0 measures that fit the needs of their local businesses, as mentioned previously.

1.5.4 The Dynamic Relationship Between Factors

As stated at the beginning of this article, the goal of the present research is to build a robust foundation for future Industry 4.0 implementation investigations. The results introduced above represent the first step of this endeavour. In this regard, the systematic review of the literature has shown that several factors have been the subjects of discussion throughout a wide range of publications, while none of the identified studies has synthesised or discussed all of the factors presented in this article. This increases the difficulty for corporations, governments, and scientist to get a holistic understanding of it. Furthermore, since most of the Industry 4.0 publications are referring to technological challenges and potentials, it may seem that these challenges, consequently, deserve more attention than others. However, as several studies have shown, not all of the challenges companies are currently experiencing are related to Industry 4.0 technologies [27]. Hence, the presented findings strengthen the idea that the factors, as well as the relationship between them, need to be discussed in greater detail. In the same vein, the findings suggest that the information on the factors themselves are sometimes contradicting, further demonstrating the need for a broader scientific discussion that helps to put the implementation of Industry 4.0 into perspective.

The three categories have been created to support that sense-making advancement by illustrating that different parties have an impact on the transitional process. For instance, the factor "Available Knowledge" has been classified as an external factor as it is mainly outside of the control of corporations. It refers to the knowledge that is available for corporations in various forms such as qualified employees on the job market and Industry 4.0 implementation guidelines. While companies can and do work together with universities and governments, it is still difficult for them to steer education policy or curriculum decisions in a particular direction. Hence, universities and governments have a more substantial impact on this factor. On the other hand, the factor "Skills Development" has been classified as an internal factor, while also mainly revolving around education as a necessary resource. Qualifying employees through programs, offering Industry 4.0 specific training, and initiating internal Industry 4.0 awareness campaigns are measures that can be taken by the corporations themselves to address the potential shortcomings of the job market. This example highlights the dynamic relationship between various factors and the created categories. In this context, the relationship between two or more chosen factors as well as between categories also depends on other factors. Going back to the education example, the ability of a company to compensate for a potential lack of qualified workers on the job market by intensifying the internal promotion of skills also depends on other factors such as whether a company has enough time and resources to increase its activities beyond the day-to-day operations. On the other hand, for governments and universities to effectively support this process, they first need to identify the needs and the specific characteristics of their local corporate infrastructure.

While it is unclear whether the presented categorisation of the factors serves its purpose or if a different and a more efficient form results from future investigations, it can be noted that relationship between the factors plays an important role and that this relationship deserves more attention. Similarly, it is too early to state that the presented list of factors is complete. The factor "Occupational Health and Safety" is a good example of this as compared to other factors; only two publications have discussed it so far in the context of the implementation of Industry 4.0. Therefore, it is likely that more factors will be added to the list over time [1].

1.6 Conclusion and Future Work

Implementing Industry 4.0 is an undertaking that depends on a number of factors that need to be considered. Furthermore, the results of the systematic review suggest that interrelations between various factors exist. These interactions offer another perspective on the implementation dynamics. Consequently, this perspective further argues in favour of a holistic approach when it comes to transitioning towards Industry 4.0. In addition, it can help to develop a broader understanding of Industry 4.0 which, for example, makes corporations and governments more aware of potential pitfalls.

1.6.1 Conclusion

Overall, the systematic literature review revealed that a number of potential factors that are important in the context of implementing Industry 4.0. However, the relevance of the identified factors cannot be derived from the findings of the review. While some factors have been mentioned more frequently than others, the question of the importance of each factor cannot be answered based on the number of references they appeared in. Furthermore, the analysis of the literature indicates that some findings contradict each other. For example, as previously mentioned, some findings suggest that the size of a company matters when it comes to the implementation of Industry 4.0 whereas other findings were not able to confirm this assumption [3, 5, 16]. Therefore, future investigations should be focused on validating the findings of the systematic literature review. As previously proposed, interviewing experts, as well as designing a survey that tests the findings of the systematic literature review, can help to create a solid foundation on which further investigations can be built. Once the most important factors are known and validated, it is essential to conduct further investigations that try to explain how companies, governmental institutions, and universities have to deal with certain factors. While knowing that the promotion of Industry 4.0 skills is important when it comes to implementing Industry 4.0, it still does not answer the question of which skills are particularly important or how certain skills should be taught.

1.6.2 Future Work

Without drifting off into trivia and without looking into the future through a crystal ball, it already can be stated that technologies referred to Industry 4.0, such as Cyber-Physical Systems and Artificial Intelligence, will have a significant impact on corporations organize work [4, 10, 22]. The answers to questions like to which degree workers will use smart glasses or how many jobs will fall victim to autonomous systems, still, cannot be answered yet [9]. Depending on which of the various interpretations and visions of Industry 4.0 paves its way to the core of how companies organize their work, the process of answering the previous questions can gradually become more than empirical curve-fitting and guesswork. What, in contrast, seems sufficiently inevitable, is that companies will have a significant share of the forces that shape the fourth industrial revolution as they, for instance, decide which machine communication protocols will be used or how the data of their employees will be processed. Against this backdrop, examining relevant factors can contribute to making the implementation of Industry 4.0 more accessible and therefore more diverse, helping those corporations that would otherwise stand on the sideline.

References

1. Badri, A., Boudreau-Trudel, B., Souissi, A.S.: Occupational health and safety in the industry 4.0 era: a cause for major concern? Saf. Sci. **109**, 403–411 (2018)
2. Bhamu, J., Sangwan, K.S.: Lean manufacturing: literature review and research issues. Int. J. Oper. Prod. Manag. (2014)
3. Braun, A., et al.: Der Weg zur Industrie 4.0-Roadmap. ZWF Z. Wirtsch. Fabr. **113**(4), 254–257 (2018)
4. Cohen, Y., Naseraldin, H., Chaudhuri, A. et al.: Assembly systems in industry 4.0 era: a road map to understand assembly 4.0. Int. J. Adv. Manuf. Technol. **105**, 4037–4054 (2019)
5. Deloitte Insights: The Fourth Industrial Revolution Is Here - Are You Ready. Deloitte Development LLC (2018)
6. Erol, S., Schumacher, A., Sihn, W.: Strategic guidance towards industry 4.0–a three-stage process model. In: International Conference on Competitive Manufacturing, vol. 9, pp. 495–501 (2016)
7. German Frank, A., Santos Dalenogare, L., Fabian Ayala, N.: Industry 4.0 technologies: implementation patterns in manufacturing companies. Int. J. Prod. Econ. **210**, 15–26 (2019)
8. Hamzeh, R., Zhong, R., Xu, X.: A survey study on industry 4.0 for New Zealand manufacturing. Procedia Manuf. **26**, 49–57 (2018)
9. Kong, X., Luo, H., Huang, G., Yang, X.: Industrial wearable system: the human-centric empowering technology in industry 4.0. J. Intell. Manuf. **30**(8), 2853–2869 (2018)
10. Li, B., Hou, B., Yu, W., Lu, X., Yang, C.: Applications of artificial intelligence in intelligent manufacturing: a review. Front. Inf. Technol. Electron. Eng. **18**(1), 86–96 (2017)
11. Liao, Y., Deschamps, F., Loures, E., Ramos, L.: Past, present and future of Industry 4.0 - a systematic literature review and research agenda proposal. Int. J. Prod. Res. **55**(12), 3609–3629 (2017)
12. Lin, D., Lee, C., Lau, H., Yang, Y.: Strategic response to industry 4.0: an empirical investigation on the Chinese automotive industry. Ind. Manag. Data Syst. **118**(3), 589–605 (2018)
13. Maghazei, O., Netland, T.: Implementation of industry 4.0 technologies: what can we learn from the past? In: IFIP International Conference on Advances in Production Management Systems, pp. 135–142 (2017)
14. McKinsey, D.: "Industry 4.0 after the initial hype", Where manufacturers are finding value and how they can best capture it (2016)
15. Moher, D. et al.: Preferred reporting items for systematic review and meta-analysis protocols (PRISMA-P) 2015 statement. Syst. Rev. **4**(1) (2015)
16. Moica, S., Ganzarain, J., Ibarra, D., Ferencz, P.: Change made in shop floor management to transform a conventional production system into an "industry 4.0": case studies in SME automotive production manufacturing. In: 2018 7th International Conference on Industrial Technology and Management (ICITM), pp. 51–56 (2018)
17. Mueller, E., Chen, X., Riedel, R.: Challenges and requirements for the application of industry 4.0: a special insight with the usage of cyber-physical system. Chin. J. Mech. Eng. **30**(5), 1050–1057 (2017)
18. Müller, J., Buliga, O., Voigt, K.: Fortune favors the prepared: how SMEs approach business model innovations in industry 4.0. Technol. Forecast. Soc. Chang. **132**, 2–17 (2018)
19. Martínez Sánchez, A., Pérez Pérez, M.: Lean indicators and manufacturing strategies. Int. J. Oper. Prod. Manag. **21**(11), 1433–1452 (2001)
20. Roblek, V., Meško, M., Krapež, A.: A complex view of industry 4.0. SAGE Open **6**(2), 215824401665398 (2016)
21. Rossini, M., Costa, F., Tortorella, G., Portioli-Staudacher, A.: The interrelation between industry 4.0 and lean production: an empirical study on European manufacturers. Int. J. Adv. Manuf. Technol. **102**(9–12), 3963–3976 (2019)
22. Saniuk, S., Saniuk, A., Cagáňová, D.: Cyber industry networks as an environment of the industry 4.0 implementation. Wirel. Netw. (2019)

23. Serdyukov, V., Serdyukova, N., Shishkina, S.: Improving operational reliability by means of artificial intelligence. Russ. Eng. Res. **38**(1), 15–18 (2018)
24. Shah, R., Ward, P.: Lean manufacturing: context, practice bundles, and performance. J. Oper. Manag. **21**(2), 129–149 (2003)
25. Sivathanu, B., Pillai, R.: Smart HR 4.0 - how industry 4.0 is disrupting HR. Hum. Resour. Manag. Int. Dig. **26**(4), 7–11 (2018)
26. Sommer, L.: Industrial revolution - industry 4.0: are German manufacturing SMEs the first victims of this revolution? J. Ind. Eng. Manag. **8**(5) (2015)
27. Stauffen, A.: Deutscher Industrie 4.0 Index. Industrie (2018)
28. Tortorella, G., Fettermann, D.: Implementation of industry 4.0 and lean production in Brazilian manufacturing companies. Int. J. Prod. Res. **56**(8), 2975–2987 (2017)
29. Tranfield, D., Denyer, D., Smart, P.: Towards a methodology for developing evidence-informed management knowledge by means of systematic review. Br. J. Manag. **14**(3), 207–222 (2003)
30. Wang, S., Wan, J., Li, D., Zhang, C.: Implementing smart factory of Industrie 4.0: an outlook. Int. J. Distrib. Sens. Netw. **12**(1), 3159805 (2016)
31. Xu, L.D., Duan, L.: Big data for cyber physical systems in industry 4.0: a survey. Enterp. Inf. Syst. 13(2), 148–169 (2019)

Chapter 2
Motivation in a Business Company Using Technology-Based Communication

Madara Pratt and Sarma Cakula

Abstract The world is experiencing many changes due to increased globalization, e-commerce and modernization. The new modernized work setting offers us many advantages and benefits; technology-based communication allows us to span time and distance among people. Technology development is moving at such a speed that social human behaviour science has not kept pace. Although there is enormous research on human behaviour, we lack new models for guiding managerial practices. Countries and companies will compete in the future on the quality of both—technology and their human resources. The question is how these two concepts can be brought together—technology and human resources? How to keep employees satisfied and motivated while the company strives to excel in technology? The aim of this research is to investigate technology-based communication's effect on the quality of relationships within the company, to determine factors affecting social relationships and also their impact on motivation and job satisfaction of employees. Artificial Intelligence (AI) and Industry 4.0 offer us enormous possibilities for advancements and improvements to these technologies. AI has a huge potential for increasing the efficiency and performance of communication technologies. The theoretical model developed by the authors is explaining how technology-based communication is affecting the quality of work relationships. An online survey was conducted, and in a sample of 102 respondents representing 26 different nationalities, it was found that the factors affecting motivation do not correlate with technology communication-related constructs, but rather are affected by management. In conclusion, it is possible to communicate through technology and motivate employees with the same efficiency as in face-to-face communication. Also, a need for new and effective communication programs and tools was recognized.

M. Pratt (✉)
Vidzeme University of Applied Sciences, Cesu 4, Valmiera, Latvia
e-mail: madara.pratt@va.lv

S. Cakula
Riga Technical University, Kronvalda 1, Riga, Latvia
e-mail: sarma.cakula@va.lv

A. Dingli et al. (eds.), *Artificial Intelligence in Industry 4.0*,
Studies in Computational Intelligence 928,
https://doi.org/10.1007/978-3-030-61045-6_2

2.1 Introduction

Increased globalization, new economy, and e-commerce has changed the ways of communication and, in turn, changed the work environment; many companies developed virtual teams with team members in different time zones, countries and even continents. This has developed new and more complex challenges for leadership because of issues arising from new communication formats [19]. The question is how to have a high performing, motivated and satisfied employees within companies where communication is fully or partly technology-based. Within this new context, a motivated workforce has been pointed out as a hallmark of competitive advantages [35]. Countries and companies will compete in the future on the quality of both—technology and their human resources [38]. Artificial Intelligence (AI) and Industry 4.0 offer us enormous possibilities for advancements and improvements to these technologies.

The traditional work relationship and forms of communication between employer and employee obviously have changed, but there is not a clear understanding of its replacement. We lack new models capable of guiding managerial behaviour in this new era of work. Steers, Wickham and Acitelli (2014) believe that it is time to redirect intellectual energies into discovering new models and research towards new models of work motivation and job performance [35]. Barrier (2001) emphasizes that companies can not continue to blindly accept and introduce components into information systems without studying the effectiveness, feasibility, and efficiency of the individual components of their information systems; these systems have to be managed [4]. Although behavioural and social scientists have done enormous research on human behaviour, researchers and developers have rarely utilized them [20]. In 2004 "ACM Conference on Computer Supported Cooperative Work" held in Chicago, the topic of how to incorporate group and organization theory in CSCW (Computer Supported Cooperative Work) was discussed [3]. It is clear from discussions that CSCW community should adopt a stronger orientation to social science disciplines [11]. AI has a huge potential for increasing the efficiency and performance of communication technologies. We can include the missing components and use an enormous amount of data and ML (Machine Learning) to improve information systems and users satisfaction. Authors have studied the efficiency and application of AI to predict employee satisfaction and turnover (attrition) rate [2, 23, 29]. ML provides us with exact and detailed calculations which are important to solve these social behaviour challenges.

The aim of this article is to investigate how technology-based communication is affecting the motivation and satisfaction of employees. Communication can be through richer medium channels as face-to-face discussions, video conferencing, phone calls and though leaner medium channels as text messages, e-mails and unaddressed documents. A literature review was conducted researching motivation and different factors affecting it within the context of communication through technology, the conceptual model developed. The survey was conducted in order to understand the technology effect on motivational factors and overall in technology use. As a result, the need for new communication technologies was recognized, and the importance of good management was emphasized.

2.2 Technology in Work Communication

Introduction of technology has brought many benefits to business companies. It allowed employees to be geographically dispersed—located in different places, countries and even continents. They can work full time, part-time and have different shifts.

Computer Supported Cooperative Work (CSCW) is a research area where the use of computing and communication technologies which support group and organizational activity are studied [11]. CSCW can be discussed in four general levels—individual, group and team, organization and industry. Family, occupation, nation, or culture are other discussed factors [11].

In order to improve task performance, increase range and speed of information, and overcome time and space distance, many companies have developed virtual teams [24]. Additional benefits of virtual teams are discussions based on knowledge, facts and improved brainstorming [22]. Virtual communication can happen using phone calls, e-mails, instant messages, video chats, videoconferences, shared screen sessions, shared files, and others. The degree of virtual communication within organization or teams can be discussed at three levels—face-to-face teams that meet personally, low virtuality teams that use synchronous, media-rich technologies and high virtuality teams that use asynchronous, media poor technologies, such as e-mail [25].

Companies can categorize their meetings in four categories by time and place [26]. STSP (same-time, same-place) are face-to-face meetings. STDP (same-time, different-place) meetings are a combination of video, audio or text. Information technology serves as a storage system. DTSP (different-time, same-place) meetings work as a storage system and are used, for example, for employees with different shifts. Within DTDP (different-time, different-place) meetings team members share the same virtual space and web [21].

Media richness theory ranks communication media according to their ability to convey complex messages. Most effective are richer medium channels as face-to-face discussions, video conferencing and phone calls and leaner medium channels as text messages, e-mails, unaddressed documents are less effective [42]. Communication within virtual teams should be developed with a variety of types of media considering working style and tasks, including richer and leaner medium channels [42].

Although technology offers many benefits, companies have to be aware of many effects it has on work design, organizational design, communication patterns and secondary social effects [15]. Communication within this context is based on knowledge and opinions, and that decreases status differences. Taking into consideration that within technology-based communication part of the intended message can be lost and, thus there is an absence of non-verbal cues, it can lead to miscommunication, lack of interpersonal relations, social rapport and less satisfied members, the amount of this communication within the company can be a significant measurement of quality of relationships among these team members [22]. It is difficult to establish and maintain mutual knowledge [8]. These misunderstandings can lead to false conclusions and assumptions about a person's character, especially when communication is across cultures [21, 39].

It was thought that increased levels of video conferencing, which is considered a rich communication media, would solve these communication problems; however, expectations were not realized [37]. Video conferencing does not require full attention as face-to-face communication and mainly matches the name to the face, but doesn't substitute for face-to-face communication, because of its inability to read non-verbal cues [14].

2.3 Quality of Work Relationships

Communication is an important factor for maintaining social relations in the company, and the lack of social information can limit social relations in the company [9]. Two of the impacts of technology-based communication are increased miscommunication and anonymity [16, 21]. In this type of communication, there is less pressure, and this affects how teams make decisions and resolve conflicts. Anonymity may have several negative effects. It results in loss of self-awareness, and people would say what otherwise they would not say in face-to-face communication. Deindividualization can result in social loafing. Having less social pressure can cause higher levels of conflicts, inability to resolve conflicts and to reach consensus in decision-making situations [21]. Miscommunication can result in conflicts because of misunderstandings and reduced communication [16]. On the other hand, there are also benefits of anonymity—better generation of ideas, greater willingness to participate and to generate more ideas [21].

For building social relations, it is suggested to have face-to-face meetings including social activities and to encourage employees to communicate informally on an everyday basis; it is normal practice to use technology for bonding social ties among employees [21]. Vignovic and Thompson (2010) suggest concentrating on developing social relations and using communication norms in order to overcome communication problems [39].

2.3.1 *Culture*

Impact of communication differs depending on culture and age differences [36]. Cultural background is an important factor influencing abilities and qualities of technology used for system developers to consider. Röse and Züblke (2001) recognized a gap in the developer's knowledge according to the culture-dependent user requirements of a foreign market [32]. Within the research about the specification, analysis, and integration of relevant intercultural variables, Röse (2004) specified, that modern and user-friendly products have to include cultural diversity as one of their key features [31]. One might think that communication problems stem from differences in linguistic codes, but Hall (1976) argues, that the problem lies in the context, which carries varying proportions of the meaning [13]. Spoken language, in fact, is an abstraction of thoughts and, in turn, written language is an abstraction of spoken language. The actual event happening is much more complex and richer than the abstraction of it within the spoken language and written language. It is natural that people choose some parts of information and unconsciously ignore others.

According to Hall (1976), any transaction can be characterized as high, low or middle context [13]. There is much more context and little or no information in High Context (HC) communications, but Low Context (LC) communication provides a lot of information with little or no hidden context. Within technology-based communication, the most complex issue is transmitting the desired message in High Context communication. As most of the meaning is not transmitted in words, it can lead to misunderstandings. To have sufficient HC communication programming has to be done; if that is not done, this communication is insufficient [13]. Controversy in LC communication can be very brief and short and not have all information explained in sufficient detail and lack emotion and personal relationships.

2.3.2 Leadership

Thompson and Coovert (2006) emphasized appropriate leadership as an important factor for improving relationships within technology-based communication [37]. Even if communication is computer-based, an undeniable part of work management is personal leadership.

Communication between the employee and his manager is critical for the company as it influences many aspects. The leadership effect on virtual teams is especially important because they have more complex challenges to overcome the structuring team process and providing task support than face-to-face teams do [19]. Because of communication restrictions, virtual teams require more active leadership. Employees have difficulties feeling like a part of the team and have limitations in building social relationships [21].

One of the theories explaining leader-employee communication is Leader-Member Exchange (LMX) theory. Graen and Cashman (1975) suggested that leaders are establishing either low-exchange relationships or high-exchange relationships with their employees [12]. This theory is focused on a dynamic relationship between leader and employee. In-group employees are the ones which leader would consider superior to other employees, more loyal, better at work, reliable. They receive more attention and benefits. Out-group members have fewer advantages; they are not trusted, and the leader doesn't involve them as much. A very significant part of this theory is communication through which leaders and members develop and maintain a beneficial exchange. Out-group members receive less attention, support, supervision, communication from a leader [7, 12, 45]. Some of the resulting consequences are turnover intent, lower job performance, overall organizational citizenship behaviour, organizational commitment and general job satisfaction [10].

2.4 Motivation

A motivated workforce has been pointed out as a hallmark of competitive advantages within a new modernized work setting [35]. Communication technology developers should consider this critical point while developing new systems. Usually, the term motivation reflects a person's internal psychological forces which move them to work [30]. Herzberg (2003) recognized the type of manager who takes direct action by the strategy "kick the person" also called KITA [17]. There are various forms of KITA—negative physical, negative psychological and positive KITA. He revealed several myths about motivation—reducing time spent at work, spiralling wages, fringe benefits, human relation training, sensitivity training, communications, two-way communication, job participation and employee counselling. Each of these approaches can be effective, but Herzberg argues KITA is not motivation and mentioned myths also are not long-term motivators [17].

Results and discussions of Hawthorne study suggest that the increase of assemblers' productivity was because they were singled out and made to feel important [43]. Results of other early research showed that people have natural striving for self-realization and that the most insistent incentive is self-respect and the respect of others [18]. Employees need to feel that they are a part of a worthwhile project, have responsibilities, and they are rewarded for outstanding performance. Money is not enough to motivate employees [18, 43]. Research in the Intermountain Health Care organization also concluded that motivating and rewarding work environment are more important than an increase in salary or benefits [41]. Research in IT sphere show that for IT professionals, the pay is not a motivator, but work itself, satisfaction, feeling of achievement are [1].

Herzberg (1959), in his research "Motivation to Work", was the first one who linked job satisfaction with motivation in his two-factor theory [18]. Hygiene (extrinsic) factors or dissatisfaction-avoidance include supervision, interpersonal relations, physical and working conditions, salary, company policies and administrative practices, benefits, and job security. When these conditions are below a level that an employee feels acceptable, he gets dissatisfied. But it is not the same in reverse—when conditions are optimal, an employee is not satisfied, but he is merely neutral. These are factors, which are essential for avoiding unpleasant situations [17, 18, 30]. Motivators (intrinsic) are factors that reward the needs of the individual to reach his aspirations and are the ones that bring actual job satisfaction. These factors are achievement, recognition for achievement, the work itself, responsibility, and growth or advancement [17, 18, 30].

Volkwein's (2003) findings confirm that positive workplace relationships and atmosphere of teamwork are the ingredients that have a highly positive impact on most satisfaction measures [40]. Interpersonal conflicts have a decreasing effect on satisfaction. Job insecurity strongly and directly influences satisfaction and increases intentions to leave. Satisfaction is influenced by a complex array of personal and situational circumstances [40].

It is very critical to keep employees motivated and have a proper communication strategy because these two factors lead to job satisfaction/dissatisfaction [28]. Job satisfaction or dissatisfaction is a result of motivation and performance [41]. Turnover intent is closely related to job satisfaction [6, 33]. Results in research in Turkey confirmed that there is a negative relationship between job satisfaction and turnover intention [44]. The cost of employee hiring and training can be very high. Authors have researched the advanced methods of excelling this problem by using ML algorithms (1–3). AI proves to bring us to the next level of computing these complex human behaviour issues.

2.5 Conceptual Model

Technology-based communication is critical because of its influence on several factors and the connections between the variables can be arranged in a conceptual model (Fig. 2.1) reflecting major constructs from a literature review [27, 28]. Herzberg's two-factor theory of motivation is the underlying theory for this model [18]. Technology-based communication is influencing Hygiene (extrinsic) factors, such as the perception of effective leadership and good relations with co-workers [28]. It is a factor for dissatisfaction-avoidance rather than becoming more satisfied. In order to have satisfied employees, the company has to not only maintain a neutral level of Hygiene (extrinsic) factors, but also maintain Motivators (intrinsic) factors,

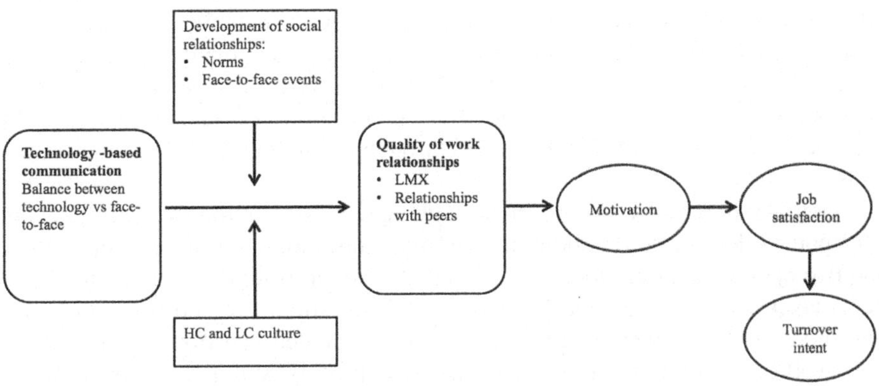

Fig. 2.1 Conceptual model of technology-based work communication (*Source* Author)

such as achievement, recognition for achievement, the work itself, responsibility, and growth or advancement [17, 18, 30].

Some of the influencing factors on motivation and job satisfaction relationship are personal characteristics, job characteristics, intrinsic and extrinsic motivation factors, commitment, job satisfaction and turnover intent [1, 34, 40, 44]. Volkwein and Zhou (2003) also added state characteristics and divided job characteristics into two parts—institutional and administrative work climate [40]. For this research, Smerek and Peterson (2007) conceptual model will be used as a base [34]. Smerek and Peterson (2007) based their model on Herzberg's two-factor theory.

Ahmed et al. (2017) extended the model; turnover intent was added as a concept influenced by job satisfaction [1]. Yücel (2012) related job satisfaction and turnover intent with a commitment [44]. Pauga and Cakula (2019) adjusted these models, and from presented literature review it can be concluded that technology-based communication (the amount of technology used) is affecting the quality of work relationships, which is mediated by culture and can be influenced by face-to-face events [27, 28]. The authors predict that more work communication is done through technology-based means, the harder it is to maintain high-quality relationships inside the company and therefore, it would influence work motivation negatively.

Quality of work relationships includes LMX and relations with peers. If the quality of work relationships is good, then it leads to higher motivation [28]. If employees are motivated, then they are satisfied, and this job satisfaction results in negative turnover intent [1, 6, 28, 33, 43].

2.6 Methodology

An online survey was conducted, covering topics of work motivation and the use of technology at the workplace. A total of 102 respondents answered the survey. There was no missing data and no outliers. The survey was conducted internationally covering 26 different nationalities and wide age range (19–63 years old). Altogether 60% ($n = 62$) of respondents have a Bachelors degree and 23% ($n = 24$) have a Postgraduate degree or higher. Gender is distributed rather equally—58% ($n = 59$) male and 40% ($n = 41$) female. In total 27.4% ($n = 28$) of the respondents communicates 100% through technology-based communication tools and 19.61% ($n = 20$) of respondents communicate 50% or less through technology-based communication tools.

The scale questions ranged from strongly agree (10) to strongly disagree (1) on a 10-point Likert scale. In addition to 93 scale questions, seven descriptive items and two open-ended questions were included. The motivation was measured by 59 items taken from Smerek and Peterson (2007) questionnaire and was adjusted to this survey [34]. Intrinsic motivation questions included questions about recognition, work itself, opportunities of advancement, professional growth opportunities, good feelings about organization and clarity of mission. Extrinsic motivators were measured with questions about senior management, effective supervision, good relationships with co-workers and satisfaction with salary. Communication technology concept was investigated using 27 Likert scale questions, including technology use at work communication, communication programs, and additional features for communication technologies. This is an initial attempt to investigate technology-based communication in the context of motivation. Various questions were added by the authors to test the possible approach and factors that could show relevance and correlation. The results were processed in PSPP program. Within this article, the most important and relevant insights from descriptive statistics and bivariate correlation will be described.

2.7 Analysis and Results

From the initial data analysis and overall descriptive measures (Table 2.1), it can be concluded that people are positive about their effectiveness of technology-based communication (mean = 7.94).

Fast Internet connection (mean = 8.74) and quality of visuals (mean = 8.02) are very important for technology-based communication. Use of images is a popular additional feature (mean = 6.21). Also, it is not uncommon to use emojis/smileys (mean = 5.22) in work communication. There is a trend of less formal communication, as emojis/smileys have been used. Respondents confirm this claim as informal (casual) communication style for work communication has a high mean (mean = 7.64). Writing long messages and long e-mails are not preferred (mean = 4.77), even though surveyed people are for many different cultures. It shows an overall tendency to use fewer words and to be more direct in communication. It could be speculated, that within the new work environment and use of technology, communication style is getting more Low Context. On top of that, many respondents use technology for informal communication with co-workers (mean = 7.64). People are not very concerned about face-to-face meetings and missing physical presence at work (mean = 4.61).

The most popular device used for work communication is computer (mean = 9.02). The tendency of using smartphones (mean = 7.56) is very high, but the tablet is a very unpopular device for work communication (3.60). E-mails (mean = 8.62) and phone calls (mean = 7.33) are the most popular means of work communication. A company chat box is also one of the communication channels that people use (mean = 5.61). Whatsapp is the most popular application for communication, but it is still rated low (mean = 5.61).

Social media channels and applications are not preferred for work communication among surveyed people—Linked-in (mean = 4.59), Telegram (mean = 4.45) and the least popular is Facebook (3.63). Cloud-based apps (ex. Google drive) seems to be popular (mean = 6.24). People use Skype (mean = 5.09) for conference calls (mean = 5.55), but it is not very commonly used communication method.

The results of this analysis show that people don't mind working using technology-based communication tools, but most preferred options for it are still e-mails and

Table 2.1 Mean comparison for technology-related constructs (N = 102)

Variable	Mean	Variable	Mean
My work communication is 100% technology-based	7.50	Computer	9.02
My work through technology-based communication tools is effective	7.94	Tablet	3.60
I use technology-based communication tools for informal communication with my co-workers	7.07	Smart Phone	7.56
Fast Internet connection is important for my work	8.74	E-mail	8.62
The quality of visuals is very important for my work	8.02	Company chat boxes	5.61
I prefer informal (casual) communication style for my work communication	7.64	WhatsApp	5.71
I like to write long and formal messages/e-mails	4.77	Telegram	4.45
I am missing the physical (face-to-face) presence at my work	4.61	Linked-in	4.59
		Linked-in	4.59
		Facebook	3.63
		Skype	5.09
		Phone (voice) calls	7.33
		Group video calls	5.55
		Cloud-based apps (ex. Google drive)	6.24
		Videos	4.04
		Images	6.21
		Stickers	4.25
		GIFs	4.07
		Emojis/smileys	5.22

phone calls. That means that developers have not offered a better solution. The provided applications seem to fail to provide all the necessary features to satisfy customers. Technology support for work communication is insufficient.

For assessing relationships between variables Bivariate (Pearson) correlation analysis was conducted. All of the scale questions showed significant correlations within their categories. Cohen's standard will be used to evaluate the correlation coefficient, where 0.50 or larger represents a strong association [5].

Table 2.2 Correlations among senior management and work itself constructs

No	Variable N = 102	1	2	3	4
1	Senior management keeps employees informed				
2	I enjoy the type of work I do	0.65			
3	My job is interesting	0.56	0.80		
4	My job gives me a sense of accomplishment	0.59	0.78	0.79	
5	I make a difference in my organization	0.60	0.60	0.62	0.70

Management communication with their employees leads to an overall feeling about the company and work itself (Table 2.2). Respondents who think their senior managers keep employees informed also think their work is interesting (0.56), enjoy the type of work they do (0.65), have a sense of accomplishment (0.59) and feel they are making a difference in their organization (0.60).

Communication leads to an overall feeling about the company. This was unexpected that employees associate work itself with senior management involvement. Employees who like their work in general also feel better about their companies (Table 2.3). Respondents who enjoy the type of work they do also feel strong sense and belonging (0.55) and commitment (0.62) to the organization. Also, they are proud to work for this organization (0.55) and care about the future of the organization (0.54). Those who feel that they are making a difference in their organization also feel a strong sense of belonging (0.62) and commitment (0.53) as well as make them care about the future of that organization (0.56).

Employees who enjoy the work they do and like the work itself have a better feeling about the organization. Results in Table 2.2 show the importance of senior management and communication within the company. It is important for employees to choose their career path wisely, but the management also can affect employee feelings about the work they do, and if they like the work they are doing.

Another benefit of effective communication is recognition, which is one of the most critical Intrinsic factors for motivation (Table 2.4). Employees feel that their good work is recognized with senior managers who keep employees informed (0.70), effectively communicate goals (0.73) and are consistent with the values of the organization (0.67). Also, employees feel their contributions are valued with senior managers who keep them informed (0.66). A supervisor has a high effect on recognition. Employees feel their good work is recognized with supervisors who manage people effectively (0.70), create an environment that creates a trust (0.71) and recognizes them for good work (0.73). Also, they feel that their contributions are valued when the supervisor is considering their ideas (0.67).

Table 2.3 Correlations among senior management and work itself constructs

No	Variable N = 102	1	2	3	4	5	6
1	I enjoy the type of work I do						
2	My job gives me a sense of accomplishment	0.78					
3	I make a difference in my organization	0.60	0.70				
4	I feel a strong sense of belonging to the organization	0.55	0.58	0.62			
5	I have a strong commitment to the organization	0.62	0.60	0.53	0.65		
6	I am proud to work for this organization	0.55	0.57	0.48	0.58	0.77	
7	I care about the future of the organization	0.54	0.57	0.56	0.61	0.69	0.77

Table 2.4 Correlations among senior management, supervisor and recognition constructs

No	Variable N = 102	1	2	3	4	5	6	7
1	Senior management keeps employees informed							
2	My employer recognizes my good work	0.70						
3	My contributions are valued by my employers	0.66	0.84					
4	My supervisor manages people effectively	0.67	0.70	0.67				
5	My supervisor creates an environment that fosters trust	0.54	0.71	0.61	0.77			
6	My supervisor recognizes me for doing good work	0.56	0.73	0.57	0.70	0.76		
7	My supervisor considers my ideas	0.52	0.67	0.51	0.61	0.74	0.75	
8	Internal candidates receive fair consideration for open positions	0.62	0.75	0.71	0.51	0.58	0.52	0.55

Senior management and supervisors are responsible for making their employees feel recognized for the work there are doing. It is clear that communication with employees is a very important factor for work motivation.

From the data analysis, it is clear that management and leadership have the main role in motivating their employees in both technology-based communication and face-to-face communication. The amount of technology-based communication within the company correlates very high with effective work through technology (0.85). Employees who work more through technology also use it for informal communication with other co-workers (0.51). Those respondents who prefer communication through technology rather than face-to-face also use these communication means for communicating with their co-workers informally (0.65). There were no significant correlations with any other constructs.

Managers have the strongest effect on motivational factors. Analysis of these results didn't show the relevance of the way communication happens, but rather on leadership skills managers are using. It shows that it is possible to communicate through technology and motivate employees with the same efficiency as in face-to-face communication. Managers have more challenges in virtual teams than those who work mostly face-to-face [28].

2.8 Conclusions

Within this article, the most relevant theory affecting work motivation was described. Herzberg's two-factor theory of motivation was reviewed, underlining the importance of Hygiene and Motivator factors within business companies. Communication technology integration in the work environment was discussed. The importance and potential of AI and Industry 4.0 have been pointed out. An updated theoretical model of technology-based communication relation with work motivation was presented in Fig. 1 [27, 28]. This model is a guide for AI systems and offers a clear structure of necessary system components which have been missing from social science research.

An online survey was conducted, and the use of current technologies was investigated. The need for new technological solutions was recognized. Even though developers have created many communication technologies, the most widely used are e-mail and phone calls. Other technologies are not offering the needed solution to replace these basic means of communication. Videoconferences and video calls are not very widely used, and it confirms Thompson and Coovert [37] findings of the efficiency of this communication tool [37]. It is very popular to use smartphones for work communication, which means that developers have to provide programs that are user-friendly for both—computers and smartphones.

In general, people are satisfied with their work efficiency using technology-based communication tools. The quality of Internet and visuals were found to be very important, and it can be predicted to gain more importance which technology developers have to keep in mind. Cloud-based programs are widely used, and this could be an increasing trend due to its practicality.

People prefer to write short messages and e-mails, use emojis/smileys and send images in their workplace communication. It shows an overall tendency to use fewer words and to be more direct in communication. This survey was conducted among people from 26 different nationalities, and different cultures were presented in the data set. It could be speculated, that within the new work environment and use of technology, general work communication style is getting more Low Context.

It was found that effective communication is key for motivational factors in the company. Senior management and employee supervisors have to inform their employees and communicate well because it is strongly affecting recognition, overall feeling about the company and liking the work they do. There was no correlation between motivational and technology-communication factors; the amount of technology used for work communication is not directly affecting motivational factors as it was expected by the authors. It can be inferred, that the management, which is communicating effectively with their employees, can do it through both—face-to-face meetings and technology-based channels. Communicating through technology-based channels is more challenging; technology developers should provide technologies that would increase the efficiency of this communication. For future research, authors recognize the need to test empirical data in a simulation model. Technological solutions for supporting company management should be created. AI has a huge potential for increasing the efficiency and performance of communication technologies. We can include the missing components and use an enormous amount of data and ML (Machine Learning) to improve information systems and users satisfaction.

Acknowledgements Technology Enhanced Learning E-ecosystem with Stochastic Interdependences - TELECI. No.1.1.1.1/16/A/154

References

1. Ahmed, S., Taskin, N., Pauleen, D.J., Parker, J.: Motivating information technology professionals: the case of New Zealand. Australas. J. Inf. Syst. **21** (2017)
2. Alao, D., Adeyemo, A.B.: Analyzing employee attrition using decision tree algorithms. Comput. Inf. Syst. Dev. Inform. Allied Res. J. **4**(1), 22–6 (2013)
3. Barley, S.R., Dutton, W.H., Kiesler, S., Resnick, P., Kraut, R.E., Yates, J.: Does CSCW need organization theory? In: Proceedings of the 2004 ACM Conference on Computer Supported Cooperative Work, pp. 122–124 (2004)
4. Barrier, T. (ed.): Human Computer Interaction Development and Management. IGI Global (2001)
5. Cohen, J.: Statistical Power Analysis for the Behavioral Sciences, 2nd edn. Lawrence Erlbaum Associates, Hillsdale (1988)
6. Coomber, B., Barriball, K.L.: Impact of job satisfaction components on intent to leave and turnover for hospital-based nurses: a review of the research literature. Int. J. Nurs. Stud. **44**(2), 297–314 (2007)
7. Dansereau Jr., F., Graen, G., Haga, W.J.: A vertical dyad linkage approach to leadership within formal organizations: a longitudinal investigation of the role making process. Organ. Behav. Hum. Perform. **13**, 46–78 (1975)
8. Driskell, J.E., Radtke, P.H., Salas, E.: Virtual teams: effects of technological mediation on team performance. Group Dyn.: Theory Res. Pract. **7**(4), 297–323 (2003)
9. Duarte, D.L., Snyder, N.T.: Mastering Virtual Teams: Strategies, Tools, and Techniques that Succeed. Wiley, Hoboken (2006)
10. Dulebohn, J.H., Bommer, W.H., Liden, R.C., Brouer, R.L., Ferris, G.R.: A meta-analysis of antecedents and consequences of leader-member exchange: integrating the past with an eye toward the future. J. Manag. **38**(6), 1715–1759 (2012)
11. Ghaoui, C. (ed.): Encyclopedia of Human Computer Interaction. IGI Global (2005)

12. Graen, G., Cashman, J.F.: A role making model of leadership in formal organizations: a developmental approach. In: Hunt, J.G., Larson, L.L. (eds.) Leadership Frontiers. Kent State University Press, Kent (1975)
13. Hall, E.T.: Beyond Culture. Anchor Press, New York (1976)
14. Hambley, L.A., O'Neill, T.A., Kline, T.J.: Virtual team leadership: the effects of leadership style and communication medium on team interaction styles and outcomes. Organ. Behav. Hum. Decis. Process. **103**(1), 1–20 (2007)
15. Hartman, K., Neuwirth, C.M., Kiesler, S., Sproull, L., Cochran, C., Palmquist, M., Zubrow, D.: Patterns of social interaction and learning to write: some effects of network technologies. Writ. Commun. **8**(1), 79–113 (1991)
16. Hertel, G., Geister, S., Konradt, U.: Managing virtual teams: a review of current empirical research. Hum. Resour. Manag. Rev. **15**(1), 69–95 (2005)
17. Herzberg, M.: One more time: how do you motivate employees? Harv. Bus. Rev. **81**(1), 87–96 (2003)
18. Herzberg, M., Mausner, B., Snyderman, B.: The Motivation to Work. Wiley, New York (1959)
19. Huang, R., Kahai, S., Jestice, R.: The contingent effects of leadership on team collaboration in virtual teams. Comput. Hum. Behav. **26**(5), 1098–1111 (2010)
20. Kraut, R.E.: Applying social psychological theory to the problems of group work. HCI Models, Theories and Frameworks: Toward a Multidisciplinary Science, pp. 325–356 (2003)
21. Levi, D.: Group Dynamics for Teams, 4th edn. Sage Publications Inc, Thousand Oaks (2014)
22. Levi, D.: Group Dynamics for Teams, 5th edn. Sage Publications Inc, Thousand Oaks (2017)
23. Loganathan, M., Ashwini, S.: A study on employee attrition and retention in manufacturing industries. Int. J. Commer. Manag. Res. **3**(7), 43–45 (2017)
24. McGrath, J., Hollingshead, A.: Groups Interacting with Technology. Sage, Thousand Oaks (1994)
25. Mesmer-Magnus, J.R., DeChurch, L.A., Jimenez-Rodriguez, M., Wildman, J., Shuffler, M.: A meta-analytic investigation of virtuality and information sharing in teams. Organ. Behav. Hum. Decis. Process. **115**(2), 214–225 (2011)
26. Mittleman, D., Briggs, R.: Communication technologies for traditional and virtual teams. In: Sundstrom, E. (ed.) Supporting Work Team Effectiveness, pp. 246–270. Jossey-Bass, San Francisco (1999)
27. Pauga, M., Cakula, S.: The impact of using technology-based communication on quality of work relationships. In: Proceedings of the International Scientific Conference - Society, Technology, Solutions, vol. 1, p. 17 (2019)
28. Pauga, M., Cakula, S.: Technology-based communication in the business company. In: Proceedings EDUCON2019 - IEEE Global Engineering Education Conference, pp. 843–850. IEEE Xplore (2019)
29. Punnoose, R., Ajit, P.: Prediction of employee turnover in organizations using machine learning algorithms. Comput. Inf. Syst. Dev. Inform. Allied Res. J. **5**(2), 22–6 (2016)
30. Reņģe, V.: Organizāciju psiholoģija. Birznieka SIA "Kamene", Riga (1999)
31. Röse, K.: The development of culture-oriented human machine systems: specification, analysis and integration of relevant intercultural variables. In: Cultural Ergonomics, pp. 61–103. Emerald Group Publishing Limited (2004)
32. Röse, K., Züblke, D.: Culture-oriented design: developers' knowledge gaps in this area. IFAC Proc. Vol. **34**(16), 11–16 (2001)
33. Rusbult, C.E., Farrell, D.: A longitudinal test of the investment model: the impact on job satisfaction, job commitment, and turnover of variations in rewards, costs, alternatives, and investments. J. Appl. Psychol. **68**(3), 429–438 (1983)
34. Smerek, R.E., Peterson, M.: Examining Herzberg's theory: improving job satisfaction among non-academic employees at a university. Res. High. Educ. **48**(2), 229–250 (2007)
35. Steers, M.L.N., Wickham, R.E., Acitelli, L.K.: Seeing everyone else's highlight reels: how facebook usage is linked to depressive symptoms. J. Soc. Clin. Psychol. **33**(8), 701–731 (2014)
36. Tannenbaum, S.I., Mathieu, J.E., Salas, E., Cohen, D.: Teams are changing: are research and practice evolving fast enough? Ind. Organ. Psychol. **5**(1), 2–24 (2012)

37. Thompson, L.F., Coovert, M.D.: Understanding and Developing Virtual Computer-Supported Cooperative Work Teams, Creating High-tech Teams: Practical Guidance on Work Performance and Technology. American Psychological Association, Washington (2006)
38. Thurow, C.: Who owns the twenty-first century? Sloan Manag. Rev. **33**(3) (1992)
39. Vignovic, J.A., Thompson, L.F.: Computer-mediated cross-cultural collaboration: attributing communication errors to the person versus the situation. J. Appl. Psychol. **95**(2) (2010)
40. Volkwein, F., Zhou, Y.: Testing a model of administrative job satisfaction. Res. High. Educ. **44**(2), 149–171 (2003)
41. Whetten D., Cameron, K.S.: Developing management skills. Prentice Hall, Upper Saddle River (2011)
42. Workman, M., Kahnweiler, W., Bommer, W.: The effects of cognitive style and media richness on commitment to telework and virtual teams. J. Vocat. Behav. Spec. Issue Technol. Careers **63**(2), 199–219 (2003)
43. Wren, D.A., Bedeian, A.G.: The evolution of management thought. Wiley, Hoboken (1994)
44. Yücel, İ.: Examining the relationships among job satisfaction, organizational commitment, and turnover intention: an empirical study. Int. J. Bus. Manag. **7**(20), 44–58 (2012)
45. Yukl, G.A.: Leadership in Organizations. Pearson, London (2010)

Chapter 3
Decision Support in Everyday Business Using Self-enforcing Networks

Christina Klüver and Jürgen Klüver

Abstract Several Artificial Intelligence-Techniques were developed and discussed in the context of Industry 4.0. There are a lot of investigations to optimize processes in manufactories and for the analysis of big data, not to name problems like automatic driving, speech and image recognition. Yet despite many achievements in these fields the daily problems of management still remain the same, which usually evade the application of formal methods. We just name three typical problems of these domains and show how these problems can be dealt with by applying a specific kind of neural networks: (a) The decision, which employees are best suited for a specific project, i.e. the problems of Human Resource Management (HRM); (b) the question, which model of procedure (MP) should be chosen when starting a new project, and finally (c) how to avoid misunderstandings when facing the problems of interdisciplinary or of virtual teams. In according models and simulations with so-called Self-Enforcing Networks we show in this article, based on empirical experience and in cooperation with different companies, that problems of these kinds can also be successfully dealt with to avoid failures in projects; because of the increasing complexity for humans in such projects the usage of according techniques could be an important improvement for dealing with problems of management. Self-Enforcing Neural Networks (SEN) allow a simple modeling of different everyday business problems, as well as the analysis and clustering of complex data or e.g. imaging diagnostics. They belong to the type of self-organized learning networks, frequently also called non-supervised learning networks.

C. Klüver (✉)
University of Duisburg-Essen, Essen, Germany
e-mail: christina.kluever@uni-due.de

J. Klüver
CoBASC Research Group, Essen, Germany
e-mail: juergen.kluever@uni-due.de

A. Dingli et al. (eds.), *Artificial Intelligence in Industry 4.0*,
Studies in Computational Intelligence 928,
https://doi.org/10.1007/978-3-030-61045-6_3

3.1 Introduction

Nowadays the concept of Artificial Intelligence (AI) became (again) most popular in every domain, using several techniques, which require a specific knowledge and experience. It became more important to understand the strengths and weaknesses of each technology [1, 2], but problems start already with the difficulty to find a common understanding of the concept [3], partially because e.g. Machine Learning (ML) and Artificial Intelligence are often used synonymously. The controversial discussions with regard to the definitions of AI and ML are not part of this contribution, but for our purpose it is important to mention that we distinguish these two domains in our research group according to the type of learning. If e.g. the models are guided by theories and if they require cognitive or rational aspects (cognitive or constructivist learning), which are important for the decisions, or to understand how humans or machines solve the given problems, then the models are assigned to AI. If the models require an amount of data (examples) and "only" the outcome is of interest (as in behavioristic learning), the models are assigned to ML. This is a pragmatic distinction between the two fields, which have much in common, especially in the used techniques. In this contribution we use a Self-Enforcing Network (SEN) and its possibilities are shown by the application to three different daily business problems, namely:

(a) the selection of suited employees for a specific project,
(b) the recommendation of the best model of procedure for project management,
(c) how to generate a common understanding to avoid misunderstandings.

According to our distinction these models are based on the understanding of the AI-concept.

3.2 Current Discussions on the Topics

Considering that we address different topics and areas of business, only a very condensed overview of these topics is given.

Human resource management (HRM) is the most common used term for the important task to manage the employees [4, 5]. Because the problems HRM has to deal with become more and more complex, the managers have to adapt their skills and react flexible according to the problems. In HRM there exist different approaches [6, 7]; the common goal is usually to bind the employees to their firms, to organize the structure of the working processes, and to offer the employees further training opportunities according to their capabilities. Meanwhile intelligent systems are proposed using several (AI-)techniques to support HRM, which means to get new possibilities for HRM but also new problems [8–10].

The digital transformation has an implication not only on HRM, which leads already to the concept of "E-HRM" (Electronic-HRM) [11, 12] or the need of

"Intelligent Human Resource Information Systems" [8], it also requires to question the practice of the project management methodology [13]:

Models or Methods of procedure (MP) should ensure that projects are successfully completed and are often not a voluntary option; in several companies, especially big ones, MP are mandatory used [14]. There exist several "classical" (e.g. sequential as Waterfall), agile (e.g. Scrum) and hybrid models (e.g. Scrumban); in addition some companies develop their own models. The challenge is to find out, which model is best suited for a concrete project. The digital transformation and the amount of data open the discussion about the necessity of new developments or changing of the methods [15].

Different AI-techniques are, e.g. in Systems Engineering, actually proposed as decision making systems [8], fuzzy neural networks and regression fuzzy models are used to forecast the effort for software development [16, 17], recurrent neural networks for software fault prediction [18] and agents are introduced to assure the quality of Rapid Application Development (RAD) [19], to name only a few. For the selection of agile development methodologies e.g. the Analytic Hierarchy Process (AHP) and supervised learning Neural Networks (NN) are proposed [20].

There exist to be sure several reasons why a project is successful or it fails; in any case the interaction between the project members and e.g. the stakeholders is an important component, which leads to the next addressed problem:

Misunderstandings in projects are discussed actually from different point of views (e.g. [21]) and the developments of frameworks or a modeling language to assure the requirements are certainly fruitful to avoid misunderstandings (e.g. [22–24]). Misunderstandings, especially because of ambiguity or different usage of the same concept, arise often in virtual and/or interdisciplinary teams [25, 26], not to mention that the interaction with AI-systems (however defined) is the next discussed challenge (e.g. [27, 28]).

In these shortly discussed areas the used techniques are very different, such as Bayesian Networks, different types of Neural Networks, Expert Systems and Fuzzy methods, Agents, Evolutionary Algorithms, Hybrid Systems etc. and the distinction between ML and AI is not always explicitly defined.

3.3 The Self-enforcing Network (SEN)

The usage of AI techniques when dealing e.g. with business problems always implies the task to understand the respective problem in terms of these techniques, in this case in terms of a specific neural network. That is, by the way, at its core a hermeneutical problem, namely of "understanding" the problem situation in the sense of hermeneutics.

A Self-Enforcing Network (SEN) is a self-organized learning neural network that we developed. Technical details for the operation of a SEN are only in the sense described to understand how SEN operates (for details e.g. [29]):

As in every neural network SEN contains of neurons, which are defined as input- and output neurons, activation functions, a (specific) learning rule and a learning rate for the changing of the weight values. The differences to most other network types are that the knowledge or the real data are represented in a so-called semantical matrix and, most important, that the weight values, which represent the connections between the neurons, are in the beginning not generated at random. The theoretical background is the assumption that a cognitive system does not learn the affiliation of an attribute (input neuron) to an object (output neuron) at random, but in a specific context. The same occurs with real data when the affiliation of the attributes to the objects is given by the concrete problem.

The learning rule is accordingly:

$$w(t+1) = w(t) + \Delta w, \textit{ and } \Delta w = c \cdot w_{ao} \tag{3.1}$$

where w is the weight value, Δw is the measure of changing the weight value, c is a constant usually defined as $0 \leq c \leq 1$, with the same purpose as the learning rate in standard neural networks, and w_{ao} the affiliation of the attribute a to the object o in the semantical matrix.

In addition, a "*cue validity factor*" (cvf) is introduced, which is a measure of how important an attribute is for the membership in each category [30]. The cvf allows the exclusion (cvf = 0.0) or dampening (0.0 < cvf < 1.0) some attributes. Equation (3.1) then becomes

$$\Delta w = c \cdot w_{ao} \cdot cvf_a \tag{3.2}$$

To perform its classifying task the SEN has to learn, like most neural networks, but here in a self-organized way. The specific learning process of SEN consists of (a) the transformation of the semantical matrix into a weight matrix, and (b) the learning runs, i.e. the enforcing, i.e. chiefly increasing, of the weight values. The result of this learning process is then accordingly given by the end activation values of those neurons that represent the specific objects.

The classification of a new object is done by the comparison of the attribute values of the new objects with the end activation values of the objects that are already part of the network. The results of this process are given to the user by different visualization algorithms, as shown in the next section.

3.4 Applications of SEN in Everyday Business

Because of the characteristics of SEN the same technique can be used for very different problems. First, we will show how e.g. a skill-matrix can be immediately used as a semantical matrix, and in consequence be directly a part of SEN. In the second model, which refers to models of procedure, the semantical matrix must contain as objects the models (or methods) of procedure (MP), which should be taken into consideration, and the according characteristics (as attributes) of each MP.

The last example is more sophisticated and is dedicated to the problem to avoid misunderstandings.

All presented models can be individually developed, which has the advantage that the models contain the specific characteristics of a company.

3.4.1 *Human Resource Management (HRM)*

In [31] we have demonstrated as a typical problem of HRM how SEN can be used in a call-center of a logistic company. The assignment of the employees to the according tasks must be done on the basis of the different skills the employees have at their disposal. The decisive question then is how to assign the employees according to their capabilities.

A very simple technique for dealing with classifying the employees with respect to their qualifications is to construct a skill matrix. A skill matrix is simply a schema where the different skills that are needed for a certain working field are assigned to the employees, according to the degree of affiliation of the skills to the single employees. It is also possible to define a "Task^{x}Skill Matrix" as proposed by [32].

This example refers to a call-center for e.g. a telecommunication company, where different *technical problems* daily arise and it is assumed that the following skills mainly play an important role to solve the requests of costumers:

Technical problem solving: knowledge about different technical devices, specialized on specific problems e.g. operating systems (code 1 to code 4).

Social skills: Handling problems in social interaction/conflict management: patiently, friendly, positive influence on costumers, competent answers.

Analytical skills: fast understanding the problem, finding quick solutions.

In addition, general technical qualification and general social skills are required.

These skills, in SEN defined as attributes, are for each employee (as objects) evaluated and inserted into a scale from 0 to 1. In total 30 employees are evaluated and accordingly labeled. The semantical matrix shows the affiliation of the skills to each employee Fig. 3.1:

SEN learns according to the learning rule the skills for each employee. We used the linear activation function and c = 0.1. In addition, the needed skills for the reason codes (the reasons for the calls) 1–4 are defined and inserted as new input vectors

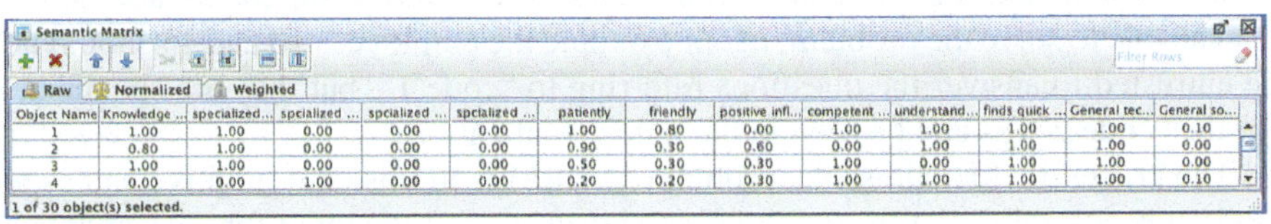
Semantic Matrix
Raw | Normalized | Weighted

Object Name	Knowledge ...	specialized...	spcialized ...	spcialized ...	spcialized ...	patiently	friendly	positive infl...	competent ...	understand...	finds quick ...	General tec...	General so...
1	1.00	1.00	0.00	0.00	0.00	1.00	0.80	0.00	1.00	1.00	1.00	1.00	0.10
2	0.80	1.00	0.00	0.00	0.00	0.90	0.30	0.60	0.00	1.00	1.00	1.00	0.00
3	1.00	1.00	0.00	0.00	0.00	0.50	0.30	0.30	1.00	0.00	1.00	1.00	0.00
4	0.00	0.00	1.00	0.00	0.00	0.20	0.20	0.30	1.00	1.00	1.00	1.00	0.10

1 of 30 object(s) selected.

Fig. 3.1 Excerpt of the semantical matrix (*Source* Author)

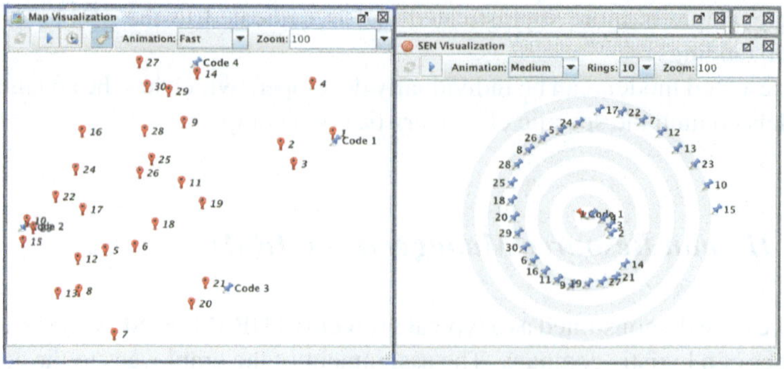

Fig. 3.2 The map-visualization on the left side (**a**) and the input centered on the right side (**b**) (*Source* Author)

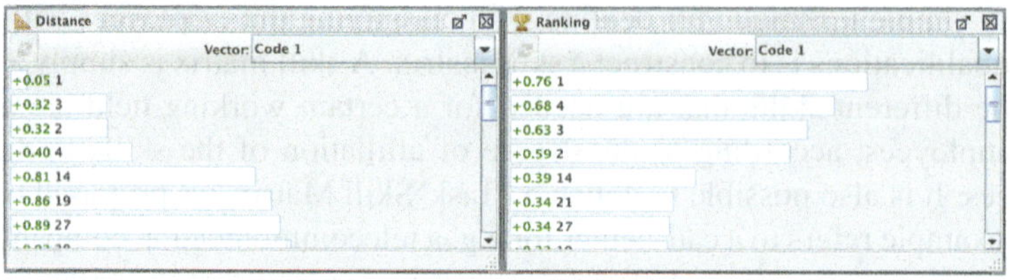

Fig. 3.3 The values of the computed distance (**a**) and ranking (**b**) (*Source* Author)

into the SEN. The goal is to assign the most competent employees to the problems a call-center has to deal with.

Two visualizations of the computed results are shown in Fig. 3.2:

The map visualization shows how the employees are attracted to the reason codes, meaning the smallest *distance* (computed as the Euclidean distance) of the employees who meet the requirements of the reason codes. The other visualization shows e.g. the reason code 1 in the middle of the center (input centered modus) and in this case those employees are attracted who have the highest activation value (*ranking*). If the concrete values are of interest, these are shown in the following visualizations (Fig. 3.3):

The two computed results show that the employee "1" is best suited for the reason "code 1". The next three others employees have the numbers 2–4. Accordingly, for each reason code can be evaluated which employees are best suited, but also who is not capable to solve the problem; in our example the employees 10 and 15 should not be entrusted to answer the questions referring to "code 1", but they are specialized for the "code 2" (as can be seen in the map visualization).

The procedure to address the problem of HRM is in this case intuitive and especially it is possible to let the employees themselves give the input for the semantical matrix. This way the employees are integrated in the process and sometimes the colleagues know much better which qualities and capabilities the employees have as e.g. a manager.

Considering that the team members have not only specific knowledge and according personalities, SEN can help to find out e.g. the best management style or the best model of procedure, as shown in the next example:

3.4.2 Models of Procedure

When a firm starts a new project it is useful and mostly necessary to structure the whole planning and performing process with a model of procedure that determines the different stages of the process, the succession of the project phases, the number of project members, the estimated time for the project and so on. There exist a lot of different models of procedure like "Scrum", "XT-model", or "Extreme Programming"; the problem for the responsible managers is to decide, which model should be used for a new project with respect to the size, time and other factors of it. We already used SEN to find a suited model for a project according to the defined criteria of the project [33].

As mentioned, some companies have mandatory requirements to use a specific model of procedure, e.g. "RUP" (rational unified process). Nowadays, when companies have to react very fast to the demands and the communication between e.g. developer and stakeholders become more important, agile models of procedures are often more suited for sub-projects because of the needs of the project or of the team members.

In this case SEN has several agile MP in the semantical matrix, which contains 60 assigned attributes for all models. The "classical" models such as RUP, Waterfall etc. are given as input vectors, interpreted as the "is-situation" in a company. The task, hence, is to show which agile MP is best suited for its combination with the standard MP. Accordingly, SEN recommends the best-suited agile procedure as shown in Fig. 3.4:

In both computed results the best-suited procedure is "DSDM" (Dynamic System Development Method), followed by "MDD" (Model Driven Development) and "FDD" (Feature Driven Development). In this recommendation, it is interesting that SCRUM is not recommended as the best one, as may be expected because of the

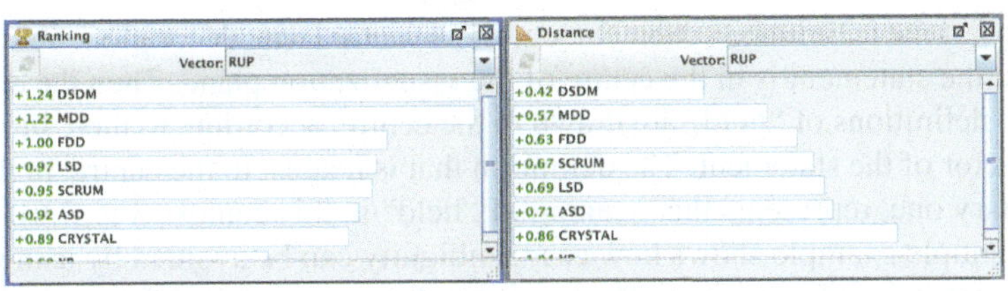

Fig. 3.4 Recommendation of SEN (*Source* Author)

popularity of the PM. The reasons are e.g. the definition of the team size (DSDM, FDD, and RUP are best suited for a large team size) and the requirement of a high experience level.

This system allows finding the best-suited combination of models and again, a company can easy define its own models and use SEN as decision support for e.g. sub-projects. Much interesting is the combination between the HRM- and PM-model if the model of procedure depends on the capabilities of the team members.

The last example is dedicated to one of the most important aspect in daily business and in projects, namely avoiding misunderstandings:

3.4.3 Misunderstandings

A. The solving of ambiguities by using a Self Enforcing Network

It is a common problem in communicative processes that many concepts that are used are ambiguous, i.e. they have more than one meaning. The concept "mouse" for example, may have the meaning "small rodent living in holes" or "component of a computer system" or even "affectionate pet name for a person one loves". As is well known ambiguities of a concept used in communication must be solved by the according context. "Context" may of course mean for example the situation in which the concept is used; it may mean an additional oral or written text that explains the meaning that the sender has in mind or it may mean non verbal explanations like, e.g. showing an object like a computer mouse in the "mouse example".

In the simulation we show "context" is used in the sense that the ambiguous concept is embedded in a certain additional text that gives the concept an unambiguous meaning.

As a suited example we chose the concept "field", which according to the Oxford Dictionary has about fifteen different meanings—for example a military one, a scientific one in mathematics and physics or a meaning with respect to sport. "He won a decoration on the field" for example refers to a military meaning of "field"; "a field is characterized by two algebraic operations" refers to a mathematical meaning of the concept. In computer science records can be divided into "fields"; in e.g. object-oriented programming means "field" the data, which are encapsulated in a class or object. Basis of the model is again the "semantical matrix"; we just show a very simple semantical matrix for "field", which can (and must) be accordingly expanded (Fig. 3.5):

We insert the mentioned statement, coded as vector (win, decoration, field), in the network; the statement is in the centre of the visualization plane. Then the complete different definitions of "field" are drawn to the centre, according to their similarities to the vector of the statement. The definition that is nearest to the centre, in this case the military one, represents the meaning of "field" in this context (Fig. 3.6).

This simple example shows how easy ambiguity can be avoided by using a SEN in an e.g. interdisciplinary team. The team members can simply define their domains and their definition of concepts.

Object Name	algebraic operation	distribution law	oil	decorations	battle	ball	winning	programming	data type
Mathematics	1.00	1.00	0.00	0.00	0.00	0.00	0.00	0.00	0.00
Science	0.00	1.00	0.00	0.00	0.00	0.00	0.00	0.00	0.00
Military	0.00	0.00	0.00	1.00	1.00	0.00	1.00	0.00	0.00
Land	0.00	0.00	1.00	0.00	0.00	0.00	0.00	0.00	0.00
Game	0.00	0.00	0.00	0.00	0.00	1.00	1.00	0.00	0.00
Computer Science	0.00	0.00	0.00	0.00	0.00	0.00	0.00	0.00	1.00
Object Oriented Programming	0.00	0.00	0.00	0.00	0.00	0.00	0.00	1.00	0.00

0 of 7 object(s) selected.

Fig. 3.5 Semantical matrix for "field" (*Source* Author)

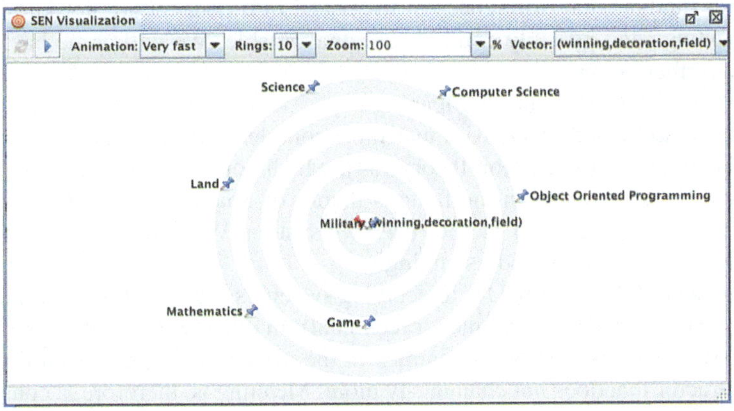

Fig. 3.6 Meaning of "field", generated by a SEN (*Source* Author)

Object Name	Business Admi...	Software Engin...	Economics	implementation
Software development	0.00	1.00	0.00	0.00
Introduction	1.00	0.00	0.00	0.00
The use of	0.00	0.00	1.00	0.00

0 of 3 object(s) selected.

Fig. 3.7 Semantic matrix for the concept "implementation" (*Source* Author)

The following real-life example shows the necessity to define the used concepts, because the ambiguity is not evident and misunderstandings occur necessarily because the employees do not know that the same concept is used very differently, as confirmed in different cases. The concept "implementation" and its understanding is kept in the semantic matrix (Fig. 3.7):

In this case the same concept is used in three different meanings and associations. In Fig. 3.8 the meaning is shown as used in Business Administration, Systems Engineering and Economics:

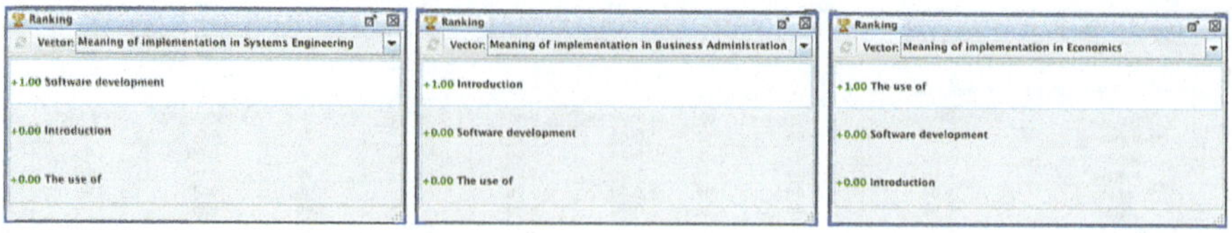

Fig. 3.8 Three different meanings for "implementation" (*Source* Author)

If such fundamental differences in the usage of a concept are not known, misunderstandings and possibly conflicts are unavoidable. We will come back to this example further below.

B. How to Generate a Common Meaning

To model the addressed problem is much more complex and the developments base on a mathematical communication theory [34], which considers not only an objective information processing as in the famous theory of Shannon and Weaver [35], but also a subjective processing by the *receiving* system. We will represent only short definitions for a common understanding:

The *meaning* of a message for a receiving system is the attractor that is generated by the signal in the system (human or technical). In other words, a message as input to the system causes a certain dynamics in the system and generates a final state, i.e. an attractor that does not change any more. Meaning is, therefore, a constructive performance of the receiving system, and a meaning for the system [36]. Other receiving systems may generate different meanings with the same signal as input; learning processes, in addition, may change the meaning for the same system.

It is a well-established methodical technique to represent the knowledge of a human in form of so-called "semantical networks" [37]. The following semantical network was constructed as a representation of the knowledge of a developer with respect to "Software" (Fig. 3.9):

This network "is" in a formal sense the developer with respect to his knowledge field "Software". If the developer receives a message that refers to the topic "Software" and that contains the concepts "implementation", "time", and "effort", then his formal model reacts in the way that these units are externally activated, i.e. they get a numerical input. The semantical network, hence, is transformed into a certain SEN, which means that the information flow between the neurons is generated via a certain "activation function", in this case a linear function.

When the network generates an attractor state its neurons are characterized by different final activation values; this "is" the meaning of the message. Figure 3.10 shows this final state where the strength of the activation values are represented by an according length of the beams:

To be sure, the meaning of such a rather short message does not consist of all neurons of the network. It is plausible to assume that only those neurons constitute the meaning of the message, which are most strongly activated. In our example the neurons effort, time and costs are also part of the network as shown in Fig. 3.9; important is that the neurons "*high*", "*several months*", and "*cost-intensive*" are the

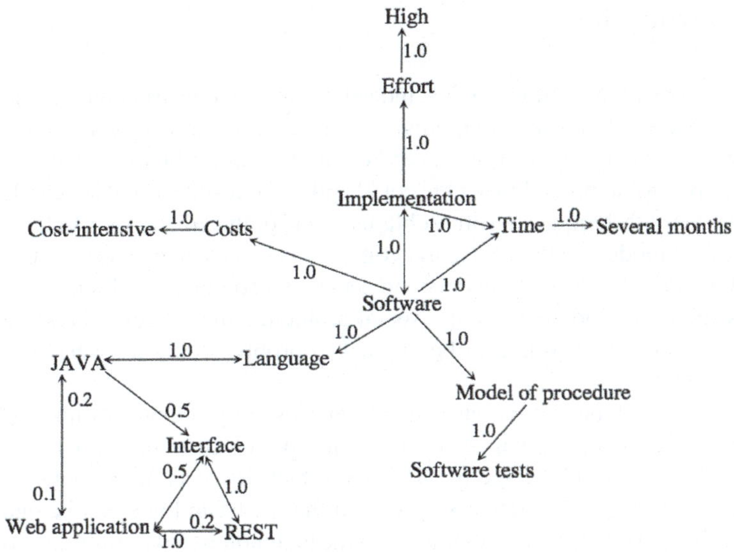

Fig. 3.9 Semantical network with the centre "Software" (*Source* Author)

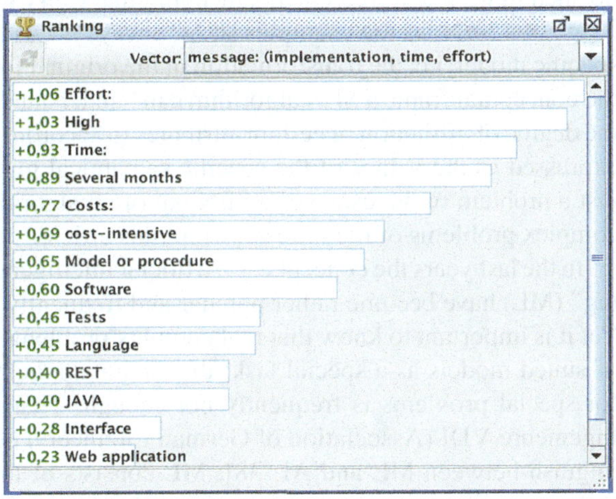

Fig. 3.10 Meaning of the message (implementation, time, effort) (*Source* Author)

accordingly activated ones. Hence the meaning of the message would be "(Effort: *high*, Time: *several months*, Costs: *cost-intensive*)". Just to mention, it is in addition possible to define a *degree of meaning*, as described in [37].

Apparently, it is possible to give an exact definition of the meaning of a message for the complex case of verbal messages. Hence a SEN system could be very usefully applied to avoid or correct misunderstandings. In addition, it is also possible to define a degree of meaning just as it is possible to define a degree of information and a degree of relevance of the message. But this cannot be demonstrated in this paper too [37].

3.5 Conclusion

The different examples show that it is indeed possible to construct mathematical theories of subjects that usually are thought a domain of the humanities or soft sciences respectively and that way to apply according models for problems usually thought to be outside of formal procedures. One could call these results as the successful applications of "hard techniques to soft problems". One of the greatest advantages of such mathematical models in the social and cognitive sciences, hence, is the obvious fact that such models enable us to understand social and cognitive problems of everyday business situations and also of other social problems. In particular the algorithm of SEN obviously is very suited to be applied to problems that are very different as regards content.

SEN was developed by us and already successfully applied to many different problems in fields like management, planning, or selecting the best new offshore locations. One great advantage of SEN is the fact that it is mathematically rather simple compared to other network types and that its basic logic can be understood rather easily by users who are not trained in the techniques and possibilities of neural networks.

Yet as the Science Fiction author Larry Niven once aptly remarked "there ain't no such thing like a free lunch", meaning that every thing has its specific cost. The usage of a SEN, as the examples show, always needs the construction of a problem specific model, i.e. the transformation of the original formulation of the problem into the conceptual frame a SEN algorithm can "understand" and deal with. For example, the degree of affiliation of certain attributes to according objects must be numerically expressed as the values of the specific semantical matrix. But that is of course not just a problem of the usage of a SEN but of all algorithms that should be applied to complex problems of the kind shown in this article.

In the last years the concepts of "Artificial Intelligence" (AI) and "Machine Learning" (ML) have become rather popular and frequently they are synonymously used. Yet it is important to know that real complex problems always need the construction of suited models as a special task; the simple application of, e.g., neural networks for special problems is frequently not enough. That is why the Verein Deutscher Ingenieure VDI (Association of German Engineers) postulated the necessity to distinguish between ML and AI [38]: ML consists of the pure application of special techniques like neural networks; AI in addition needs the construction of specific models to capture the complexity of problems like those described in this article. SEN is a very suited algorithm to construct AI models, but also applicable for ML tasks.

References

1. Davenport, T.H., Ronanki, R.: Artificial intelligence for the real world don't start with moon shots. Harv. Bus. Rev. 108–116 (2019)
2. Plastino, E., Purdy, M.: Game changing value from artificial intelligence: eight strategies. Strat. Leadersh. **46**(1), 16–22 (2018). https://doi.org/10.1108/SL-11-2017-0106
3. Kühl, N., Goutier, M., Hirt, R., Satzger, G.: Machine learning in artificial intelligence towards a common understanding. In: Proceedings of the 52nd Hawaii international conference on system sciences, pp. 5236–5245 (2019). http://www.hdl.handle.net/10125/59960
4. Armstrong, M., Taylor, S.: Amstrong's Handbook of Human Resource Management Practice, 13th edn. Kogan Page, London (2014)
5. Hovhannisyan, S., Machado, C.F.: Mutual relationship of human resource management and technology. In: Melo, P.N., Machado, C. (eds.) Management and Technological Challenges in the Digital Age London, pp. 197–210. CRC Press Taylor & Francis Group, Boca Raton (2018)
6. Jantan, H., Hamdan, A.R., Othman, Z.A.: Intelligent techniques for decision support system in human resource management. In: Devlin, G. (Ed) Decision Support Systems Advances (2010). http://www.intechopen.com/books/decision-support-systems-advances-in/intelligent-techniques-for-decision-support-system-in-human-resource-management. Accessed 10 Jun 2020
7. Lee, H.W., Pak, J., Kim, S., Li, L.-Z.: Effects of human resource management systems on employee proactivity. J. Manag. **45**(2), 819–846 (2019)
8. Masum, A.-K., Beh, L.-S., Azad, A.-K., Hoque, K.: Intelligent human resource information system (i-HRIS). A holistic decision support framework for HR excellence. Int. Arab. J. Inf. Technol. **15**(1), 121–130 (2018)
9. Parry, E., Battista, V.: The impact of emerging technologies on work: a review of the evidence and implications for the human resource function. Emerald Open Res. **15** (2019). https://doi.org/10.12688/emeraldopenres.12907.1
10. Cohen, T.: How to leverage artificial intelligence to meet your diversity goals. Strat. HR Rev. (2019). https://doi.org/10.1108/SHR-12-2018-0105
11. Bhatta, N.M.K., Thite, M.: Agile approach to e-HRM project management. In: Thite, M. (ed.) e-HRM Digital Approaches Directions & Applications, pp. 57–72. Routledge, New York (2019)
12. Ziebell, R.-C., Albors-Garrigos, J., Schultz, M., Schoeneberg, K.P., Perello-Marin, M.R.: eHR cloud transformation implementation approach and success factors. Int. J. Intell. Inf. Technol. **151**, 1–21 (2019)
13. Janssens, J.: Digital transformation. Journeys in a digitized reality. Advanced Methodologies and Technologies in Business Operations and Management. IGI Global, pp. 282–294 (2019). https://doi.org/10.4018/978-1-5225-7362-3ch021
14. Nissen, V.: Development and handling of procedure models for the selection and implementation of business software. In: Nissen, V. (ed.) Advances in Consulting Research Recent Findings and Practical Cases, pp. 133–146. Springer Nature Switzerland, Cham (2019)
15. Papadakis, E., Tsironis, L.: Hybrid methods and practices associated with agile methods, method tailoring and delivery of projects in a non-software context. Procedia Comput. Sci. **138**, 739–746 (2018)
16. Vitor de Campos, P., Souza, P., Guimaraes, A.J., Souza Araujo, V., Rezende, T.S., Silva Araujo, V.J.: Regularized fuzzy neural networks to aid effort forecasting in the construction and software development. **9**(6), 13–26 (2018). arXiv:1812.01351
17. Nassif, A.B., Azzeh, M., Idri, A., Abran, A.: Software development effort estimation using regression fuzzy models. Comput. Intell. Neurosci. (Hindawi) 1–17 (2019)
18. Turabieh, H., Mafarja, M., Li, X.: Iterated feature selection algorithms with layered recurrent neural network for software fault prediction. Expert Syst. Appl. **122**, 27–42 (2019)
19. Fatima, F., Javed, M., Amjad, F., Khan, G.U.: An approach to enhance quality of the RAD model using agents. J. Am. Sci. **14**(9), 47–55 (2018)

20. Sharma, A., Bawa, R.K.: A roadmap for agility estimation and method selection for secure agile development using AHP and ANN. In: Satapathy, S. Bhateja, V., Raju, K., Janakiramaiah, B. (eds.) Data Engineering and Intelligent Computing Advances in Intelligent Systems and Computing, vol 542, pp. 237–245. Springer, Singapore (2018)
21. Stingl, V., Geraldi, J.: Errors lies and misunderstandings. Systematic review on behavioural decision making in projects. Int. J. Proj. Manag. **35**, 121–135 (2017)
22. Vizcaíno, A., García, F., Piattini, M., Beecham, S.: A validated ontology for global software development. Comput. Stand. Interfaces 66–78 (2016)
23. Walker, D.H.T., Davis, P.R., Stevenson, A.: Coping with uncertainty and ambiguity through team collaboration in infrastructure projects. Int. J. Proj. Manag. **35**, 180–190 (2017)
24. Schön, E.-M., Thomaschewski, J., Escalona, M.J., Koch, N.: A modeling language for agile requirements engineering. In: International Conference on Information Systems Development (ISD 2018) (2018)
25. Caschera, M.C., Ferri, F., Grifoni, P.: From modal to multimodal ambiguities a classification approach. J. Next Gener. Inf. Technol. (JNIT) **4**(5) (2013). https://doi.org/10.4156/jnit.vol4.issue5.10
26. Bracken, L.J., Oughton, E.A.: 'What do you mean?' The importance of language in developing interdisciplinary research. Trans. Inst. Br. Geogr. **31**(3), 371–382 (2006). https://doi.org/10.1111/j.1475-5661.2006.00218.x
27. Riedl, M.O.: Human-centered artificial intelligence and machine learning. **1**(1), 33–36 (2019). arXiv:1901.11184, https://doi.org/10.1002/hbe2.117
28. Dellermann, D., Calma, A., Lipusch, N., Weber, T., Weigel, S. Ebel, P.: The future of human-AI collaboration. A taxonomy of design knowledge for hybrid intelligence systems. In: Hawaii International Conference on System Sciences (HICSS), Hawaii, USA, pp. 274–283 (2019)
29. Klüver, C.: Self-enforcing networks (SEN) for the development of (medical) diagnosis systems. In: International Joint Conference on Neural Networks (IJCNN). Proceedings of the IEEE World Congress on Computational Intelligence (IEEE WCCI) 2016, Vancouver, pp. 503–510 (2016)
30. Klüver, C.: Steering clustering of medical data in a self-enforcing network (SEN) with a cue validity factor. In: IEEE Symposium Series on Computational Intelligence, Athens, pp. 1–8 (2016). https://doi.org/10.1109/SSCI.2016.7849883
31. Klüver, C., Klüver, J.: Nature analogous methods for everybody. Applications to human resource management. In: Brito, A.C., Tavares, J.M.R.S., De Olivera, C.B. (eds.) Proceedings of the 2014 European Simulation and Modelling Conference (ESM'2014) Eurosis-ETI, pp. 171–177 (2014)
32. Hayashiguchi, E., Endou, O., Impagliazzo, J.: The "i competency dictionary" framework for IT engineering education. In: IEEE World Engineering Education Conference (EDUNINE) (2018). https://doi.org/10.1109/EDUNINE.2018.8450945
33. Klüver, C.: Solving problems of project management with a self enforcing network (SEN). In: Klüver, C., Klüver, J. (eds.) Social-cognitive complexity computational models and theoretical frames special issue of CMOT. Comput. Math. Organ. Theory (Springer) **18**(2), 175–192 (2012)
34. Klüver, J., Klüver, C.: On Communication. An Interdisciplinary and Mathematical Approach. Springer, Dordrecht (2007)
35. Shannon, C.E., Weaver, W.: The Mathematical Theory of Information. University of Illinois Press, Urbana (1949)
36. Klüver, J.: Communication and understanding mathematical foundations and practical application. In: Klüver, C., Klüver, J. (eds.) Social-cognitive complexity computational models and theoretical frames special issue of CMOT. Comput. Math. Organ. Theory (Springer) **18**(2), 211–231 (2012)
37. Klüver, J., Klüver, C.: Social Understanding. On Hermeneutics Geometrical Models and Artificial Intelligence. Springer, Dordrecht (2011)
38. Heizmann, M., Braun, A., Hüttel, M., Klüver, C., Marquardt, E., Overdick, M., Ulrich, M.: Artificial intelligence with neural networks in optical measurement and inspection systems. At – Automatisierungstechnik (De Gruyter) **68**(6), 477–487 (2020). https://doi.org/10.1515/auto-2020-0006

Chapter 4
Approaches for Cognitive Assistance in Industry 4.0 Assembly

Thomas Schack, Benjamin Strenge, and Kai Essig

Abstract The high degree of customization and small lot sizes necessitate a considerable amount of manual assembly work in industry 4.0 scenarios. Cognitive assistance systems could enable workers with limited experience to fulfill these roles. Approaches for designing such systems are illustrated by ADAMAAS-I ("Adaptive and Mobile Action Assistance for Industry"), an advanced cognitive assistance system using smart glasses in the context of industry 4.0 production scenarios. The aim of ADAMAAS-I is to move from stationary systems to a mobile and adaptive action support and monitoring system that reacts dynamically to human actions and prevent mistakes by giving appropriate assistance. For this purpose, ADAMAAS-I uses 3D in-situ instructions via augmented reality and state-of-the-art machine learning techniques for action recognition. Furthermore, ADAMAAS-I incorporates a systematic and largely automatized method from cognitive science for analyzing users' action representations in long-term memory. This information about individual memory structures is used to anticipate human errors and to improve the accuracy of action recognition. By these means, ADAMAAS-I provides individualized, user- and situation-specific assistance, e.g. for manual assembly tasks in an industrial context.

4.1 Introduction

A primary goal of "industry 4.0" visions is the production of highly customized items with lot size one. Many current endeavors focus on automatizing production in so-called smart factories with highly flexible self-controlled cyber-physical systems,

T. Schack (✉) · B. Strenge
Neurocognition and Action Research Group, Center for Cognitive Interaction Technology (CITEC), Bielefeld University, Bielefeld, Germany
e-mail: thomas.schack@uni-bielefeld.de

B. Strenge
e-mail: benjamin.strenge@uni-bielefeld.de

K. Essig
Rhine-Waal University of Applied Sciences, Kamp-Lintfort, Germany
e-mail: kai.essig@hochschule-rhein-waal.de

A. Dingli et al. (eds.), *Artificial Intelligence in Industry 4.0*,
Studies in Computational Intelligence 928,
https://doi.org/10.1007/978-3-030-61045-6_4

i.e. embedded systems interconnected through the internet of things. Nevertheless, a substantial amount of manual work by humans complementing automatized processes will remain necessary in the foreseeable future [8, 9]. Due to the significant changes of the work environment and processes in upcoming smart factories, the nature of this human work will likely differ from current jobs, too. Specifically, personnel of all degrees of qualification may need to adapt to completely new tasks, leading to the requirement of lifelong learning.

Obviously, these foreseeable changes and issues substantiate an extensive demand for appropriate cognitive assistance systems that suitably support and facilitate learning processes. The ADAMAAS-I ("Adaptive and Mobile Action Assistance for Industry") system prototype, which was investigated at Bielefeld University, Germany, demonstrates innovative approaches for this purpose. As a successor of the ADAMAAS system [6, 7] it consists of a cognitive tool for measuring users' memory structures, an action recognition system, and 3D in-situ augmented reality (AR) instructions provided by intelligent smart glasses.

This chapter provides an overview about specific ethics-informed worker-centered system design approaches and characteristic system features, focusing on how information about task-related memory structures can be used to tailor the action recognition and feedback components to specific users and activities.

4.2 Designing Cognitive Assistance in an Ethical and Worker-Centered Manner

In most cases, complex technical systems for cognitive assistance in industry 4.0 will need to record, process, and sometimes store a lot of sensitive information about their users and their activities. This obviously generates privacy-related and potentially also corresponding legal issues. Furthermore, it is clear that cognitive assistance systems may also introduce a broad range of other ethical issues, e.g. when working processes are changed by their usage, or workers build dependencies on these systems. All of these ethical aspects may not only be worrying due to their mere existence, but also because they may frustrate successful introduction and application of the systems due to a lack of user acceptance.

In line with well-known user-centered and agile development methodologies, the "Agile Worth-Oriented Systems Engineering" approach (AWOSE, Fig. 4.1) was designed to aid project managers, engineers and stakeholders to incorporate ethical analyses in system development [18]. In a preparatory phase, a rough description of the system vision (e.g. using a concept draft or illustrative mockup) and stakeholder models, e.g. personas [12], serve as the basis for identifying potential ethical issues with reference to the "model for the ethical evaluation of socio-technical arrangements" (MEESTAR) [11]. The MEESTAR requires analyzing ethical implications of using a system along three axes: The impacted stakeholders (referencing to specific personas, organizations, and societies), the ethical "dimensions" (e.g. safety,

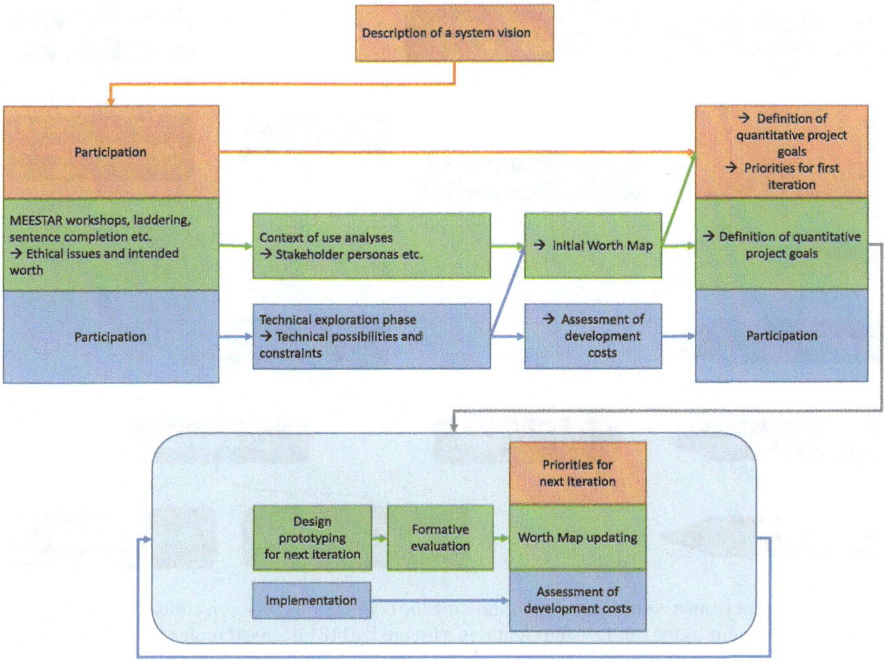

Fig. 4.1 The AWOSE process model. (Image by [18] licensed under CC BY 4.0, http://creativecommons.org/licenses/by/4.0)

privacy, or autonomy), and the expected severity. This analysis is done in interdisciplinary workshops by a diverse set of direct and indirect system stakeholders. After identification, the severity of each ethical issue is assessed based on a 4-point ordinal scale that ranges from "*completely harmless*" to "*should be opposed from an ethical viewpoint*". The ethical issues can then be associated to the corresponding system components, features, and usage scenarios via custom variants of worth map diagrams (Fig. 4.2) as used in worth-centred development [5]. This explicit modeling of the connection between system components and their wanted and unwanted implications facilitates decision making in each cycle of iterative agile development.

The AWOSE process model assigns the respective activities to three different roles, each of which could be fulfilled by more than one actual person: The "customer" (orange boxes in Fig. 4.1), the "worth designer" (which is similar to usability engineers or user experience specialists, but with a focus on ethical issues and intended positive outcomes of system use; green boxes in Fig. 4.1), and the "developer" (blue boxes in Fig. 4.1). The overall goal of this approach is to make customers or system owners aware of the impact of their choices and assist them in making better informed decisions regarding feature prioritization and selection among different possible design alternatives.

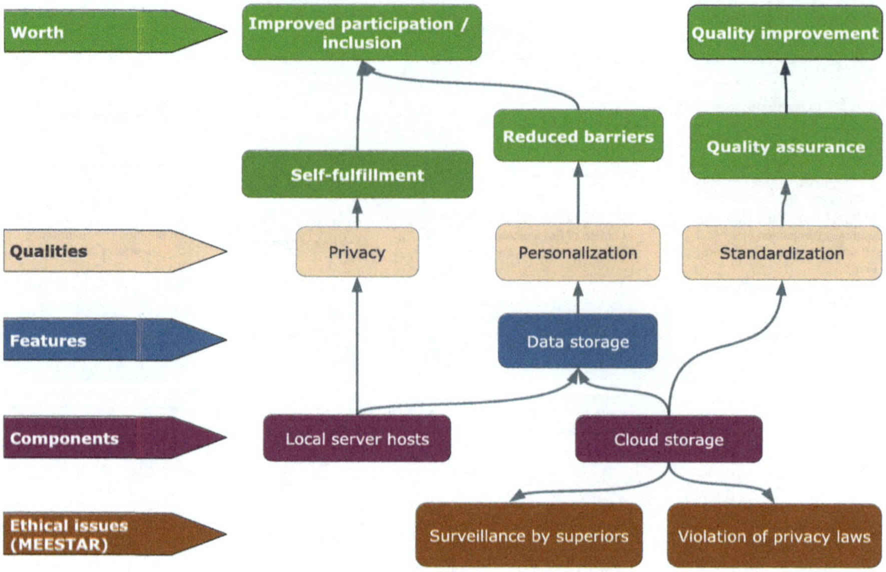

Fig. 4.2 Excerpt from a worth map sketch showing the connections between ethical issues, intended outcomes of system usage, and system features. (Image by [18] licensed under CC BY 4.0, http://creativecommons.org/licenses/by/4.0)

4.3 Task-Related Human Memory Assessment

In order to provide appropriate cognitive assistance for action sequences, a system needs information when and which assistance is actually required. The only alternatives are (1) to assist every single step, which may hamper learning and/or cause unwanted distraction, depending on the respective user's previous knowledge, and (2) to not provide any assistance by default and demand that users explicitly ask for it, which requires conscious attention to interaction with the system. Both alternatives prevent smooth and seamless task execution and additionally deplete users' limited cognitive resources.

In order to extenuate these problems, ADAMAAS-I uses an approach based on the "structural-dimensional analysis of mental representations" (SDA-M) method [15], which captures human memory structures with a semi-automatized task-related survey procedure. Dedicated software tools for this purpose, such as the "QSplit SDA-M Suite" (Fig. 4.3), show representations of each pair of task-related actions to users before system usage and ask whether they think that these actions are sequentially associated during task execution. The SDA-M method then yields data about the strengths of associations between chunks related to action representations in each user's memory.

Fed with this data, ADAMAAS-I estimates the probability that a specific user will (intend to) execute an action a_j after successfully executing a given task-related

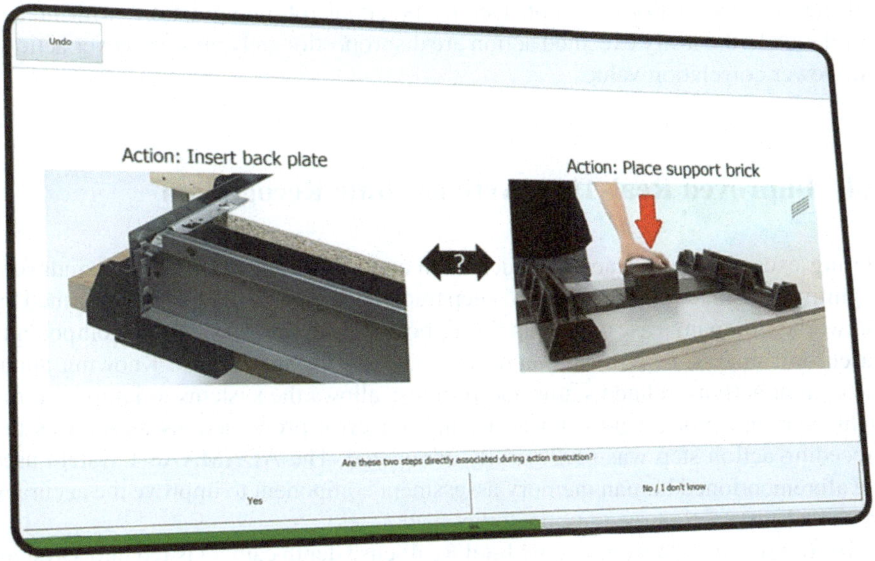

Fig. 4.3 The QSplit SDA-M tool user interface for human memory assessment showing two action steps for a manual assembly task (*Source* Author)

action a_i by assessing their mental representation structures using the "Correct Action Selection Probability Analysis" (CASPA) algorithm [19]. Based on a cognitive model derived from the ACT-R theory [1], CASPA applies a soft-max function to the activation values of actions, which in turn are functions of the pairwise correlations r_{a_k,a_l} between actions as retrieved by SDA-M.

Let A be the set of all task-related actions, $C_S \subset A$ the subset of all correct follow-up actions in a given situation, a_i the last action that has just been executed by the worker, a_{i-1} the previous action before a_i, and s a "noise factor" related to the assumed cognitive arousal of the worker. Then P_S estimates the probability that the worker would know what to do in the current situation without assistance:

$$P_S = \sum\nolimits_{a_c \in C_S,\ r_{a_i,a_c} > 0} \frac{e^{\rho(a_i,a_c)/s}}{\sum\nolimits_{a_x \in A \setminus \{a_{i-1},a_i\},\ r_{a_i,a_x} > 0} e^{\rho(a_i,a_x)/s}} \tag{4.1}$$

with

$$\text{Activation } \rho(a_k, a_l) := \log\left(\frac{r_{a_k,a_l}}{1 - r_{a_k,a_l}}\right) \tag{4.2}$$

Basically, this calculation yields a higher probability of success when one or more correct follow-up actions are strongly associated to the previously executed action and the incorrect actions are not. Hereby potential follow-up actions with higher correlation to the lastly executed action are disproportionately preferred over actions with lower correlation values.

4.4 Improved Real-Time Activity State Recognition

During execution of an activity such as manual assembly processes in industry, cognitive assistance systems need to keep track of the actions that were performed by the worker. In many cases, this can be supported by action recognition components based on cameras, sensor data, and machine learning techniques. Knowing about the current activity-related status and progress allows the systems to jump in at the right time and proactively offer assistance for error-prone actions as soon as the preceding action step was successfully completed. The ADAMAAS-I system uses the aforementioned human memory assessment component to improve the accuracy of its real-time action recognition component.

RGBD sensor data from a static Intel RealSense depth camera is fed into a system of convolutional neural networks (CNNs) that detects local hand-object interactions [17] and the global state of the current activity [16]. The current activity state during assembly is then tracked by a probabilistic state machine that is updated via Bayesian inference combining the frame-wise state likelihood output of the CNN system with SDA-M-based estimations of user-specific action transition probabilities [16]. For this purpose the SDA-M-based CASPA algorithm had to be modified to generate a worker-specific matrix of state transition probabilities. In deviation from standard calculations the value of the SDA-M analogon for correlation of an action feature vector with itself was not set to $r_{a_i,a_i} = 1$ but to $r_{a_i,a_i} = 1 - \epsilon$ with the constant ϵ chosen such that any well-established task-related mental representation structures approximate the expected probability of staying at an arbitrary action from one frame to the next based on the typical duration that domain experts need for an action in relation to the camera frame rate.

A preliminary evaluation using over 12,000 test frames from $N = 26$ participants in a manual assembly scenario in industry indicated that this approach could significantly (Fisher's exact test: $p < 0.001$) improve the accuracy of action recognition from 83.43% using only the plain CNN classifier to 85.02% correct classifications when informed by worker-specific SDA-M data.

4.5 User-Specific 3D In-Situ Augmented Reality Assistance

As soon as the ADAMAAS-I activity state recognition system noticed that the user (almost) completed a task-related action, its SDA-M-based task-related memory component assesses whether the user will likely know what to do or need assistance for the next action step. In the latter case, augmented reality smartglasses can show

Fig. 4.4 In-situ 3D instruction indicating where to install a product part (green) (*Source* Author)

static or animated 3D in-situ instructions directly at the place where the user needs to act (Fig. 4.4). This in-situ augmentation with task-related instructions has been shown to be superior to side-by-side display layouts that require cognitive mapping between instructions and the actual relevant positions in space [3]. Due to the limited field of view of currently available smart glasses hardware, ADAMAAS-I smoothly leads the user's gaze towards this place with the "SWave" attention guidance system [13] to avoid that workers overlook or need to "search" their surrounding for assisting instructions.

4.6 Evaluation

An early research prototype of the 3D in-situ augmented reality system for manual assembly assistance was tested with $N = 28$ employees at Hettich, one of the world's leading manufacturers of furniture fittings. Participants included both professional carpenters and joiners with extensive experience in manual assembly, as well as clerks and office workers with limited task-related experience. Participants were shown an instruction video and then asked to assemble a specific drawer system mockup. Whenever they attempted to execute a wrong action during the assembly the ADAMAAS-I system prototype showed 3D in-situ instructions for the current step, which were displayed by Microsoft HoloLens smartglasses. The System Usability Scale (SUS) [4] yielded a total score of 73.2, which is interpreted as a "good" [2] and above-average result [14]. The learnability factor [10] reached a score of 76.8 points.

Additionally, a preliminary user test with $N = 6$ participants that had mild cognitive or psychological impairments was conducted to test the generalizability of results to special populations. The test task, manual assembly of wooden birdhouses, was carried out at a sheltered workshop of a local diaconal organization. For this scenario, an even higher SUS score of 85.8 was reached, which is interpreted as an "excellent" result [2]. Qualitative user statements were universally positive as well, including the system was "*cool and fun to use*", "*good and helpful*", and "*overall great*".

It should be stressed again though that these preliminary evaluations had relatively small sample sizes and used an early research prototype of the ADAMAAS-I system that was not yet mature enough for productive usage. The results should therefore be interpreted with great caution, since long-term effects and influences of different usage contexts are unexplored so far.

4.7 Conclusion and Future Work

Manual assembly by human workers will remain an irreplaceable component to reach the industry 4.0 goals of extensive customization and extremely small lot sizes. Well-grounded ethical analyses and a systematic integration of their results into development processes of technical assistance systems are mandatory to establish proper user acceptance and avoid pitfalls related to legal issues, especially with respect to privacy protection.

In order to assist workers in their new tasks, advanced cognitive assistance systems such as ADAMAAS-I could use task-related memory analyses to anticipate human mistakes during action sequences in industrial scenarios and prevent them by providing augmented reality instructions for problematic action steps. Additionally, information about users' memory structures can improve state-of-the-art action recognition techniques, helping the system to detect slips and intervene when users unexpectedly act incorrectly. Furthermore, cognitive assistance systems must also aim at reducing the amount of unnecessarily displayed instructions in order to avoid distraction during work. As a result, fewer conscious system interactions shall be required. Since users can update the information about their task-related memory structures e.g. by repeating an SDA-M assessment procedure from time to time, cognitive assistance systems may also follow users' learning curves and adjust the anticipative instruction system accordingly. All of these aspects may have positive influence on learning processes, which shall help workers fulfill their various new duties in smart factories and other workplaces during the transition to industry 4.0.

Recent research (B. Strenge et al., unpublished) has indicated that SDA-M-based prediction of individual error likelihoods and task-related expertise could not only be applied to manual assembly in industrial scenarios but also to functional movement sequences and even learning of strategic actions for competitive settings. Further scientific and applied development work shall assess in how far these approaches scale to more complex real-world tasks in modern industry, how environmental factors influence their applicability, and which sensory modalities and system interaction designs are most suited for providing helpful feedback to workers in different industrial settings such as manufacturing and logistics.

Acknowledgements Many thanks go to Dr. Matthias Schröder (Apple Inc.) for his extensive support in evaluating the integration of SDA-M-based data into his CNN action recognition system. Furthermore we would like to thank Oleg Strogan for his seminal work on the QSplit SDA-M software, as well as Patrick Renner and Jonas Blattgerste for developing AR components for the ADAMAAS-I system prototype.

References

1. Anderson, J.R., Bothell, D., Byrne, M.D., Douglass, S., Lebiere, C., Qin, Y.: An integrated theory of the mind. Psychol. Rev. **111**(4), 1036–1060 (2004)
2. Bangor, A., Kortum, P., Miller, J.: Determining what individual SUS scores mean: adding an adjective rating scale. J. Usability Stud. **4**(3), 114–123 (2009)
3. Blattgerste, J., Renner, P., Strenge, B., Pfeiffer, T.: In-situ instructions exceed side-by-side instructions in augmented reality assisted assembly. In: Proceedings of the 11th PErvasive Technologies Related to Assistive Environments Conference, PETRA'18, pp. 133–140. ACM, New York, NY, USA (2018)
4. Brooke, J.: SUS: a 'quick and dirty' usability scale. In: Usability Evaluation in Industry, pp. 189–194. Taylor and Francis, London (1996)
5. Cockton, G.: Designing worth is worth designing. In: Proceedings of the 4th Nordic Conference on Human-Computer Interaction Changing Roles - NordiCHI'06. ACM Press (2006)
6. Essig, K., Strenge, B., Schack, T.: ADAMAAS: towards smart glasses for mobile and personalized action assistance. In: Proceedings of the 9th ACM International Conference on PErvasive Technologies Related to Assistive Environments, PETRA'16, pp. 46:1–46:4. ACM, New York, NY, USA (2016)
7. Essig, K., Strenge, B., Schack, T.: The superglasses-cognitive technologies as task assisting devices for sport and rehabilitation. In: Gangyan, S., Cruz, J., Jaenes, J.C. (eds.) Sport Psychology. Linking Theory to Practice. Proceedings of the XIV ISSP World Congress of Sport Psychology, pp. 258–259. International Society of Sport Psychology (2017)
8. Harbour, R., Scemama, S.: Surprise: robots aren't replacing humans in key areas of manufacturing (2017). https://www.forbes.com/sites/oliverwyman/2017/02/03/surprise-the-correct-answer-is-not-always-to-go-with-the-robot-just-ask-some-automakers/
9. Kaeser, J.: Why robots will improve manufacturing jobs (2017). https://time.com/4940374/joe-kaeser-siemens-robots-jobs/
10. Lewis, J.R., Sauro, J.: The factor structure of the system usability scale. In: Proceedings of the 1st International Conference on Human Centered Design: Held as Part of HCI International 2009, HCD 09, p. 94–103. Springer, Berlin (2009)
11. Manzeschke, A., Weber, K., Rother, E., Fangerau, H.: Ethical questions in the area of age appropriate assisting systems. German Federal Ministry of Education and Research, VDI/VDE Innovation+ Technik GmbH (2015)
12. Pruitt, J., Adlin, T.: The Persona Lifecycle: Keeping People in Mind Throughout Product Design. Elsevier, Amsterdam (2006)
13. Renner, P., Pfeiffer, T.: Attention guiding techniques using peripheral vision and eye tracking for feedback in augmented-reality-based assistance systems. In: 2017 IEEE Symposium on 3D User Interfaces (3DUI). IEEE (2017)

14. Sauro, J.: Measuring usability with the system usability scale (SUS) (2011). https://measuringu.com/sus/
15. Schack, T.: Measuring mental representations. 203–214 (2012)
16. Schröder, M., Ritter, H.: Deep learning for action recognition in augmented reality assistance systems. In: ACM SIGGRAPH 2017 Posters on - SIGGRAPH'17. ACM Press (2017)
17. Schröder, M., Ritter, H.: Hand-object interaction detection with fully convolutional networks. In: 2017 IEEE Conference on Computer Vision and Pattern Recognition Workshops, CVPR Workshops, Honolulu, HI, USA, July 21–26, 2017, pp. 1236–1243 (2017)
18. Strenge, B., Schack, T.: AWOSE - a process model for incorporating ethical analyses in agile systems engineering. Sci. Eng. Ethics (2019)
19. Strenge, B., Vogel, L., Schack, T.: Computational assessment of long-term memory structures from SDA-M related to action sequences. PLOS ONE **14**(2), 1–19 (2019)

Part II
Modelling for Industry 4.0

We cannot wait until there are massive dislocations in our society to prepare for the Fourth Industrial Revolution.

Robert J. Shiller
Yale University

Chapter 5
Modelling and Simulation in Industry 4.0

William de Paula Ferreira, Fabiano Armellini, Luis Antonio de Santa-Eulalia, and Claudia Rebolledo

Abstract The advancement of Industry 4.0 and its enabling technologies (e.g., the Internet of Things, Cyber-Physical Systems, and Big Data analytics) pose new challenges to the field of simulation due to the increasing complexity of systems to be modelled. Despite several advances in this domain, the literature still lacks a state-of-the-art review of simulation-based approaches in the Fourth Industrial Revolution. This paper therefore aims to provide a systematic literature review of peer-reviewed publications between 2011 and 2018 in the field of modelling and simulation applied in Industry 4.0, with a focus on manufacturing and logistics systems. This review contributes to the literature in two significant ways. First, it provides a broad coverage of current publications on the theme by reviewing 112 papers, using a meta-analysis approach. Moreover, it presents a detailed evaluation of different elements from these publications, including research activities (application area, methods, frameworks and software tools), gaps and directions. Around 80% of the papers analysed were published in the last two years. Finally, the results suggest that the main simulation approaches used in Industry 4.0 are agent-based or mixed methods and identify Digital Twin (DT) as the most promising simulation technology.

W. de Paula Ferreira (✉) · F. Armellini
Department of Mathematics and Industrial Engineering, Polytechnique Montreal, 2900 boul. Edouard-Montpetit, Montreal, QC H3C 1J4, Canada
e-mail: william.ferreira@polymtl.ca

F. Armellini
e-mail: fabiano.armellini@polymtl.ca

L. A. de Santa-Eulalia
Université de Sherbrooke, 2500 boul. de l'Université, Sherbrooke, QC J1K2R1, Canada
e-mail: l.santa-eulalia@usherbrooke.ca

C. Rebolledo
HEC Montréal, 3000 chemin de la Côte-Sainte-Catherine, Montreal, QC H3T 2A7, Canada
e-mail: claudia.rebolledo@hec.ca

A. Dingli et al. (eds.), *Artificial Intelligence in Industry 4.0*,
Studies in Computational Intelligence 928,
https://doi.org/10.1007/978-3-030-61045-6_5

5.1 Introduction

The term Industry 4.0 first emerged at the Hannover Fair in 2011 [1]. Although there is not yet a common definition, it can be considered a new socio-technical phenomenon, based on the intensive integration of digital information communication technologies (ICT), which can promote disruptive changes in the interactions between people, organisations and technologies [2]. In industry 4.0, "traditional industry boundaries will vanish due to the reorganisation of value-creating processes and cause severe changes within and across organisations" [3, p. 24].

Industry 4.0 can also be defined based on its requirements and enabling technologies [4]. Overall, a recent study [5] identified 12 design principles within the scope of industry 4.0, namely: horizontal integration; vertical integration; real-time capability; virtualization; decentralization; modularity; interoperability; smart factory; smart product; service orientation; corporate social responsibility; and product personalization. The same study also identifies 14 technology associated with the Industry 4.0: Simulation; Cyber-Physical Systems (CPS); Semantic Technologies; Internet of Things (IoT); Internet of Service (IoS); Internet of People (IoP); Cloud Computing; Big Data Analytics; Blockchain; Cybersecurity; Augmented Reality; Automation and Industrial Robots; and Additive Manufacturing.

A seminal reference in Industry 4.0 is the final report of the Industrie 4.0 working group [4], which first described the vision, potential, research requirements and priority areas for actions. According to this report, Industry 4.0 has the potential to meet individual customer requirements, optimize decision making, increase resource productivity and efficiency, and create value opportunities through new services. To that effect, the group identified five research requirements: (1) horizontal integration through value networks; (2) end-to-end engineering across the entire value chain; (3) vertical integration and networked manufacturing systems; (4) new social infrastructure in the workplace; (5) cyber-physical systems technology, and eight priority areas for action: (1) standardization and reference architecture; (2) managing complex systems; (3) a comprehensive broadband infrastructure for industry; (4) safety and security; (5) work organisation and design; (6) training and continuing professional development; (7) regulatory framework; and (8) resource efficiency.

Several Industry 4.0 initiatives have already been launched worldwide, usually through bi- or tripartite collaboration (e.g., university-industry-government). Even though some publications report several ongoing projects [1], Industry 4.0 is still in its infancy, and most examples are either in the planning stage or are pilot projects. The vertical and horizontal integration of IT systems and managing complex systems remain major issues for companies [1, 4, 6]. Furthermore, there is still a lack of research about the risks, costs, revenue potential and implementation barriers of Industry 4.0, as well as a lack of support and accompaniment to companies wishing to move towards this new social-technical paradigm [3]. In this scenario, modelling and simulation is seen as a primary technology in Industry 4.0 [4, 5, 7].

The application of simulation technologies is being incorporated as part of industry leaders' strategy for implementing Industry 4.0, such as General Electric, Siemens and Airbus [8]. As reported in [8], examples of industrial applications include the use of simulation by Siemens for systems planning, operation, and maintenance; the application of simulation by GE for asset management and optimization; and the employment of simulation by Airbus to monitor and optimize production processes.

Simulation is a research method that has been extensively applied throughout the supply chain of different business sectors [9–12]. It is also considered one of the most relevant techniques in the field of operations management (OM) [10, 13]. In manufacturing and logistics, simulation is considered a set of technological tools and methods that allows the experimentation and validation of products, processes, systems design and configuration, and system performance prediction. It also supports education, training and decision making, helping to reduce costs and development cycles [11]. However, the advancement of Industry 4.0 and its enabling technologies (e.g., IoT, Big Data analytics) pose new challenges to the field of simulation, mainly due to the increasing complexity of systems to be modelled.

Although there are several reviews on simulation, they are not in the context of Industry 4.0 [11] and either do not review the literature in a systematic way [14] or limit its scope to a specific aspect or simulation technique [15]. Also, none of these reviews uses meta-analytic approaches. Thus, this paper aims to provide a systematic literature review and meta-analysis of peer-reviewed publications on simulation in the context of Industry 4.0, published between 2011 and 2018, with a focus on manufacturing and logistics systems. This work is intended to contribute to the literature in two significant ways: first, through providing wider coverage of the specialized literature on the topic by reviewing 112 papers using a meta-analytic approach; and second, by summarizing current research activities (areas of application, methods, frameworks and software tools) and identifying research gaps, trends and opportunities.

The main research question (RQ) this paper seeks to answer is: RQ#1.0—What is the current state of the art of simulation-based approaches for manufacturing and logistics systems in the context of the Industry 4.0? The following sub-questions will guide the research process: RQ#1.1—What are the most influential studies, key journals and geographic distribution, and how has publication evolved over time? RQ#1.2—What are the studies' research methods, approach, application fields and main contributions? RQ#1.3—What are the essential features, modeling and simulation approaches, frameworks and software tools employed? RQ#1.4—What are the current research efforts, gaps and main directions?

This research paper is structured as follows: after introducing the problematic, relevance, aims and research questions, it presents the research method and design in Sect. 5.2. Section 5.3 presents the results from both quantitative and qualitative analysis, which are followed by the discussion in Sect. 5.4. Finally, the conclusions, limitations and some future research directions are addressed in Sect. 5.5.

5.2 Methodology

This research follows the Preferred Reporting Items for Systematic Review and Meta-Analysis (PRISMA) [16], described next.

5.2.1 Sampling and Data Collection

The data collection follows an iterative two-phase process. The search protocol used in the first iteration of the identification phase was built in three steps. First, the electronic data sources Web of Science Core Collection (WoS) and Scopus were selected. Second, the search string was constructed. The electronic search strategy for one data source, the Web of Science Core Collection, is shown next: (TS = (Simulation*) AND TS = ("Industry 4.0" OR "Industrie 4.0" OR "4th industrial revolution" OR "Fourth Industrial Revolution"). Third, three eligibility criteria were applied: (1) Date of coverage: the search ranges from 2011, with the emergence of the term "Industry 4.0" [1], to October 19, 2018; (2) Document types: only journal articles were included at this stage as they were considered more reliable due to the rigor of the evaluation process; (3) Language: English. Finally, the search protocol was applied in each data source, and papers containing the terms in either the titles, abstracts, or keywords were retrieved (WoS = 72; Scopus = 108) and combined in one single database, resulting in a total of 180 papers.

Even though the initial sample covers journals that publish the most prominent, high-quality, proceedings papers, such as Procedia CIRP, IFAC-PapersOnLine and Procedia Manufacturing, in a second iteration, the search protocol was slightly changed to include all types of documents, with the exception of journal articles, and applied in each data source. The ones with at least three citations (39 documents) were selected and included for further analysis, resulting in a sample of 219 documents.

For the screening phase, the first step consisted in removing the duplicates, using EndNote X7 and double-checking references. As a result, the sample size was reduced from 219 to 181 papers. The titles and abstracts were then analysed and the inclusion and exclusion criteria E1and E2 set out in Table 5.1 were applied, reducing the sample from 181 to 113 papers. In this stage, the 62 papers excluded due to E1 occurred mainly because they focused on other areas (e.g., energy, healthcare, construction) rather than on manufacturing or logistics systems. Next, the backward and forward snowball sampling technique [17] was applied in the 10 most cited references matching the inclusion criteria in order to verify whether any relevant references had been left out of the initial sample. Twenty additional papers were identified. The full-text articles were then assessed for eligibility and E3 was applied. A total of 112 papers were included in the sample. A third researcher, an expert in the field, double-checked the reference lists (articles excluded and included in the sample) to reduce potential bias. The sampling process is summarized in Fig. 5.1.

Table 5.1 Inclusion and exclusion criteria

Criteria	Description	Occurrences
Exclusion	E1: Does not fit with the scope of the research	68
	E2: Full text could not be assessed or is not in English	4
	E3: TERM[a] is used only as a keyword, example fact or cited expression	13
Inclusion	Simulation-based approaches are part of the main research effort and are directly related to Industry 4.0 or its main enabling technologies	112

[a]TERM refers to: Industry 4.0 or Simulation

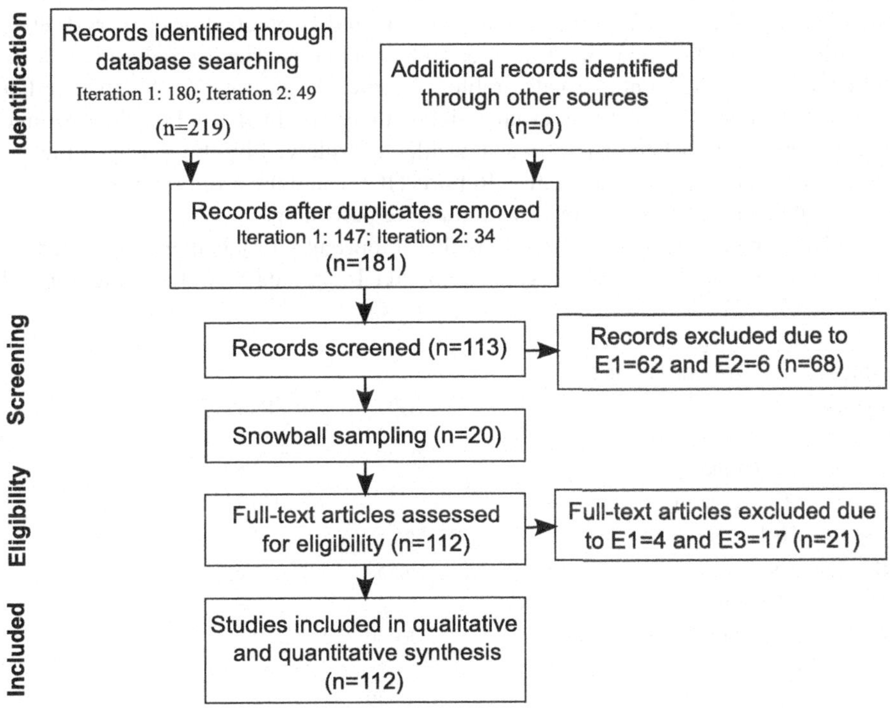

Fig. 5.1 PRISMA flow chart (*Source* Author)

5.2.2 Data Analysis

This research uses a mixed-method approach, combining both qualitative and quantitative methods. The data analysis was divided into two stages: first, we performed the meta-analysis (quantitative synthesis) through graphical and tabular methods of descriptive statistics, mainly to answer RQ#1.1. Then, after developing an initial classification scheme (coding), we conducted the content analysis.

5.3 Results

This section present the main results of the quantitative and qualitative analysis.

5.3.1 *Meta-analysis*

The final sample, used for both meta-analysis and qualitative analysis, is composed of 112 papers, that is, 101 journal articles (90.2%) and 11 proceedings papers (9.8%), according to the classification of the electronic databases. The journal articles were published in 56 different journals. Table 5.2 indicates the top 10, which together represent nearly 50% of the journal articles analysed. In this table, the absolute number (n) of published articles was considered, followed by the journal's impact factor (IF), based on Journal Citation Reports (JCR), and the proportion of papers in the sample from this journal and the cumulative percentage.

At least one article was published in other important journals in the field, such as Proceedings of the IEEE, Robotics and Computer-Integrated Manufacturing, Journal

Table 5.2 Top 10 journals with IF

Journal	n	IF	Proportion (%)	Cumulative (%)
IFAC-PapersOnLine	14	*	12.5	12.5
Procedia Manufacturing	7	*	6.2	18.7
Procedia CIRP	6	*	5.3	24.0
IEEE Access	4	3.557	3.6	27.6
Manufacturing Letters	4	*	3.6	31.2
IEEE Transactions on Industrial Informatics	3	5.430	2.7	33.9
Computers in Industry	3	2.850	2.7	33.6
International Journal of Production Research	3	2.623	2.7	39.3
International Journal of Production Research	3	2.623	2.7	39.3
Int. J. Adv. Manuf. Technol.	3	2.601	2.7	42.0
Sensors	3	2.475	2.7	44.7

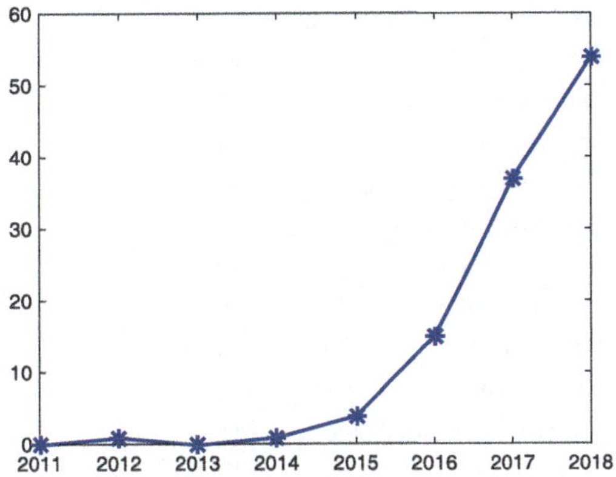

Fig. 5.2 Publications per year (*Source* Author)

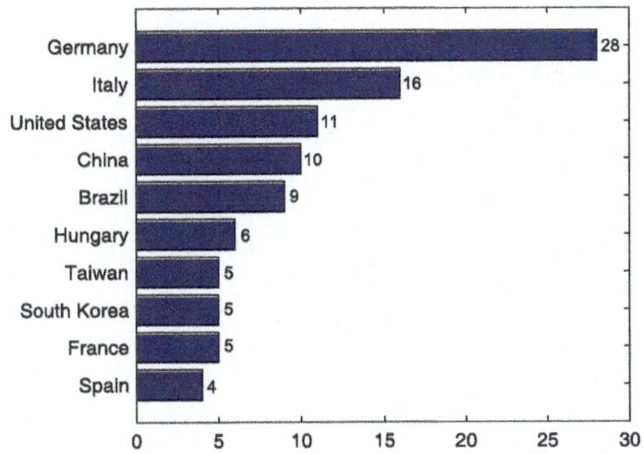

Fig. 5.3 Publications per country (*Source* Author)

of Manufacturing Systems, Journal of Manufacturing Technology Management, Simulation Modelling Practice and Theory, and International Journal of Simulation Modelling. Furthermore, analyzing the proceedings papers and the primary source of the articles published in the journals IFAC-PapersOnLine, Procedia CIRP and Procedia Manufacturing, we found three main conference organisers: the International Federation of Automatic Control (IFAC), with 14 papers, followed by the College International pour la Recherche en Productique (CIRP), with 6 articles, and the Institute of Electrical and Electronics Engineers (IEEE), with 6 publications.

Figure 5.2 shows how publications linking simulation-based approaches with Industry 4.0 evolved over time. Around 80% of the papers analysed were published in the last two years. The geographical distribution of the publications is exhibited in Fig. 5.3. The top 10 most frequent were considered. Twenty other countries (e.g., Canada, UK) are also represented in the sample with up to three papers. We considered the institutions of all authors. However, few papers (9.8%) involved authors or institutions from different countries. Looking at the distribution by continent, we find: Europe (73 papers; 65%); Asia (21 papers; 19%); North America (12 papers; 11%); and South America (9 papers; 8%).

Table 5.3 Keywords frequency

Keyword	n	%
Simulation	72	64
Industry 4.0	54	48
Cyber-physical systems	36	32
Embedded systems	21	19
Internet of things	19	17
Decision making	16	14
Real time	13	12
Digital twin	11	10
Discrete event simulation	11	10
Multi-agent systems	10	9

Table 5.4 Top 15 most cited studies

Reference	Cited
Derler et al. [18]	306
Wang et al. [19]	70
Rosen et al. [20]	60
Hehenberger et al. [21]	39
Mosterman and Zander [22]	34
Tao et al. [23]	28
Uhlemann et al. [24]	24
Schleich et al. [25]	23
Turner et al. [26]	23
Tao and Zhang [27]	21
Fortino et al. [28]	18
Uhlemann et al. [29]	15
Qi and Tao [30]	14
Penas et al. [31]	12
Weyer et al. [32]	11

Table 5.3 presents the most frequent keywords, including two used in the query. This analysis is important to initially identify the main constructs addressed in the content of the studies. In total, 1,200 keywords were collected from the 112 studies. The top 15 most cited papers included in the sample, considering the total number of citations in the Scopus database, are listed in Table 5.4.

Figure 5.4 exhibits the bibliometric network, considering the bibliographic coupling among the 112 publications. This is an important visualization technique that gives a direct indication of the relatedness of the studies [33]. The edges (lines) represent the bibliometric coupling relations. The nodes represent the publications,

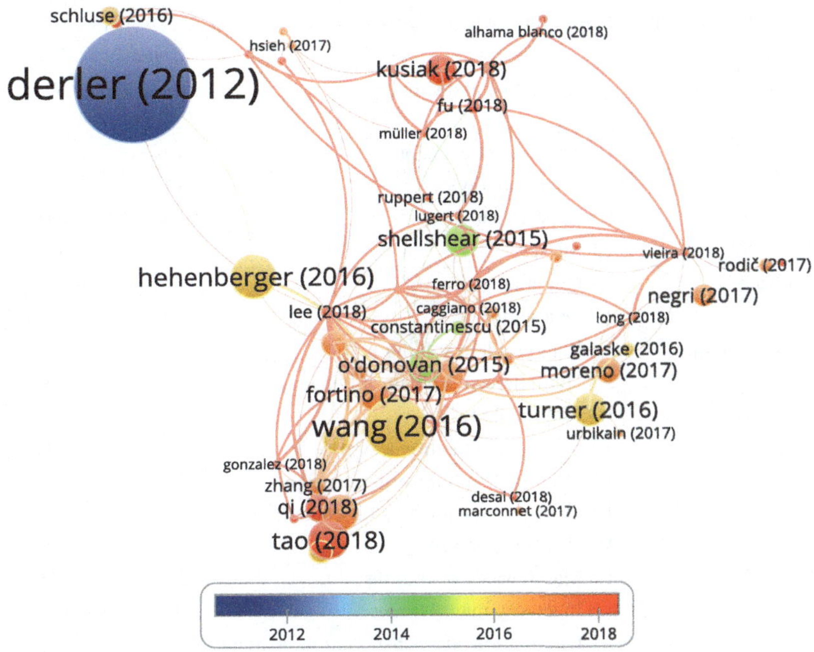

Fig. 5.4 Bibliometric network—bibliographic coupling (*Source* Author)

scaled in terms of the number of times the publication was cited, and its color represents how publications are distributed over the years. The image was produced with VOSviewer® software. Although the journal articles were published in 56 journals, a strong bibliographic coupling among publications was found, indicating an overlay in the reference lists of the articles. As indicated by Van Eck and Waltman [33, p. 287], "the larger the number of references two publications have in common, the stronger the bibliographic coupling relation between the publications."

5.3.2 *Content Analysis*

First, a classification scheme with five categories, as presented in Table 5.5, was defined to assess the main aspects of the studies and guide the content analysis. The first and second categories seek to identify the research method and approach adopted in the studies. Complementarily, the third category, taken from Jahangirian et al. [10], is used to assess their empirical nature. Category four, adapted from Negahban and Smith [11] and from Tako and Robinson [34], is used to classify the studies per area of application. Category five, adapted from Jahangirian et al. [10], is used to identify the main simulation methods employed. The key results from the content analysis were then computed using the defined classification scheme,

Table 5.5 Key information from the content analysis

Categories		Counts	%
1. Research method	1.1 Modelling and simulation	67	59.8
	1.2 Case study	23	20.5
	1.3 Theoretical-conceptual	19	17.0
	1.4 Literature review	15	13.4
	1.5 Experiment	10	8.9
	1.6 Survey	3	3.6
2. Approach	2.1 Quantitative	62	55.4
	2.2 Qualitative	50	44.6
3. Empirical nature	3.1 Real problem solving	18	16.1
	3.2 Hypothetical problem solving	44	39.3
	3.3 Methodological	50	44.6
4. Area of application	4.1 System and facility design	15	13.4
	4.2 Material handling system design	10	8.9
	4.3 Production planning and scheduling	15	13.4
	4.4 Real-time control	15	13.4
	4.5 Maintenance operations	6	5.4
	4.6 Inventory management	0	0
	4.7 Distribution and transportation planning	2	1.8
	4.8 Supply chain	6	5.4
5. Simulation methods	5.1 Discrete event simulation (DES)	24	21.4
	5.2 System dynamics (DS)	7	6.3
	5.3 Agent-based simulation (ABS)	26	23.2
	5.4 Mixed methods	39	34.8

summarized in Table 5.5, where each publication could be classified in no, one or multiple categories. Other methods and techniques identified were computed and analysed separately.

The majority of the studies (67 papers) used modelling and simulation as the research method, representing 59.8% of the sample. Additionally, 19 theoretical-conceptual papers (16.9%) and 15 review papers (13.4%) were identified, without overlapping. In general, the review papers focused on a specific simulation method and/or Industry 4.0 enabling technology (5—Digital Twin; 3—Discrete Event Simulation; 1—Distributed Simulation; 1—Big Data analytics; 2—CPS; 1—Virtual Reality; 1—Additive manufacturing; 1—smart manufacturing). It is important to point out that these reviews have a very different scope from our research. Moreover, 23 papers were classified as case studies, 10 as experimental research and 3 used the survey method. Concerning the studies' approach, 62 papers (55.4%) used quantitative methods, and 50 papers (44.6%) used qualitative approaches. As regards their empirical nature, 23 papers (20.5%) aimed to solve a real problem using real data. Another 44 papers (39.3%) used hypothetical problems or did not use real data. The majority, 50 papers (44.6%), used a methodological approach, with no application or experimental study.

Most of the studies (49%) focus on system and facility design, production planning and scheduling, real-time control and material handling system design. Other areas like maintenance operations, distribution and transportation planning and supply chain management were the focus of 12.5% of the papers. As concerns simulation methods, most studies used mixed/hybrid methods (39 papers), mainly combining the Discrete-Event Simulation (DES), Agent-Based Simulation (ABS) and System Dynamics (SD). However, we also find other methods and combinations, such as DES with Virtual Reality (VR) and DES with Computer-Aided Technologies (CAx), the latter comprising CAD/CAE/CAM as well as Finite Element Analysis (FEA). While 23.2% used ABS, 21.4% used DES, and 6.3% used SD. Digital Twin (DT), considered to be a new simulation paradigm [14], is the main unit of analysis of 18 papers (16.1%). Additionally, 22 papers (19.6%) used CAx technologies, 16 papers (14.3%) Big Data analytics, 11 papers (9.8%) Meta-heuristics [35], 9 papers (8.0%) Mathematical Programming (MP), 5 papers (4.5%) Petri Net Simulation, and 3 papers (2.7%) Artificial Neural Network.

A total of 40 different frameworks were identified, primarily respecting Cyber-Physical Production Systems (CPPS), Smart Factories or Digital Twins. These frameworks are usually built by the combination of the following Industry 4.0 technologies: Automation and Industrial Robotics, IoT, Big Data, Cloud Computing, and some Artificial Intelligence techniques. One example can be seen in Wang et al. [19], which proposed a smart factory framework that integrates industrial network, cloud computing, supervisory control and smart objects (SO), characterized by a self-organized multi-agent system (MAS) with feedback and coordination based on big data analytics. Five papers proposed frameworks for modelling and simulating IoT systems, an example can be seen in Fortino et al. [28], which also modelled IoT systems in terms of MAS. The analysed studies also used very different software tools. Of these, the most frequent in the sample are Tecnomatix Plant Simulation®, used in 10 papers (8.9%), and MATLAB/Simulink, used in 6 papers (5.4%).

5.4 Discussion

For the discussion, we return to the research questions proposed in the introductory part of the paper.

RQ#1.1—What are the most influential studies, key journals and geographic distribution, and how has publication evolved over time?

The most influential studies are listed in Table 5.2 and the key journals in Table 5.2. Although there are some specialized scientific journals that cover modelling and simulation, the articles analysed were published in 52 different journals. Despite this, the bibliometric coupling analysis showed that the articles are closed related. The geographic distribution of publications is displayed in Fig. 5.3, with a higher concentration in Europe, in accordance with Liao et al. [1]. Very few papers involved authors or institutions from different countries, suggesting that there remains considerable room for international collaboration. Modelling and simulation is an Industry 4.0 enabling technology [4, 5]. However, most studies about modelling and simulation in the context of the Fourth Industrial Revolution (around 80%) were just published in the last two years, as it emerged as a new research area.

RQ#1.2—What are the studies' research methods, approach, application fields and main contributions?

Most papers used modelling and simulation as their research methods. However, a small number of case studies, controlled experiments and empirical surveys were identified. Almost 30% are theoretical-conceptual or review papers. About 50% are more qualitative in approach, while 55.4% are quantitative. The studies' application area of simulation concentrates on system and facility design, production planning and scheduling, real-time control, or material handling system design. However, only 20.5% aimed to solve a real problem or used real data. The main contribution of 50% of the papers relies on the proposition of new theoretical-conceptual frameworks or reference models. Moreover, some papers made a practical contribution by developing cases studies, such as [36], which provided a CPPS architecture framework and implementation in an actual piston-manufacturing plant, or experiments, as in [22]. Even though we partially assessed the empirical nature of the studies using the scheme proposed by [10], most of them used case studies or controlled experiments as validation techniques for their proposed artifact (simulation framework).

Nonetheless, we identified three empirical surveys that aimed to assess the importance and use of modelling and simulation in Industry 4.0 context in practice. For example, in [37], the authors conducted two surveys in 515 South Korean small and medium enterprises (SME) in the manufacturing sector, one to quantify the utilization and effects of modelling and simulation in South Korea and the other to identify the practical effects of applying modelling and simulation in real manufacturing processes. They found that even though modelling and simulation have the potential to increase companies' operational performance, it has not been intensely used by South Korean manufacturing SMEs, mainly due to the high costs of the software tools, hardware and lack of experts in the field. In [38], the authors conducted an empirical survey with 170 lean production experts from different domains to evaluate the current status of value stream mapping (VSM) and address its future sustainability in the context of the Fourth Industrial Revolution. They found that there is a need for a combination of lean practices and Industry 4.0.

RQ#1.3—What are the essential features, simulation approaches, frameworks and software tools employed?

The most significant portion of the studies used multiple simulation methods and techniques, mainly combining DES, ABS and SD with other technologies in different ways to develop more realistic planning and exploratory models to manage complex systems that cope with essential features of Industry 4.0 as described in [5], such as: interoperability, decentralization, modularity, virtualization, real-time capabilities, and vertical and horizontal integration. While most of the proposed frameworks usually link modelling and simulation with some specific Industry 4.0 enabling technology, such as Cyber-Physical Systems (CPS), Internet of Things (IoT), Big Data analytics, and Cloud Computing,, the papers about Digital Twin are those that seem to combine a higher variety of modelling and simulation methods, techniques and Industry 4.0 enabling technologies. This seems the most promising modelling and simulation technology to converge manufacturing and logistics and physical and virtual worlds. Furthermore, over 13 software tools were identified and often more than one software tool was used. Tecnomatix Plant Simulation®, developed by Siemens, and MATLAB/Simulink from MathWorks appeared more frequently. However, there are many other software tools available on the market. An introductory guide for hybrid simulation modelers with a software tools list can be found in [39]. Furthermore, a complete and detailed software tool list for agent-based modelling and simulation is presented in [40].

RQ#1.4—What are the current research efforts, gaps and main directions?

The main research efforts can be summarized in four points: first, researchers are developing new simulation-based approaches (e.g., Digital Twin, Co-Simulation) to cope with the multiple features and technologies of Industry 4.0. Second, they are developing new simulation frameworks by combining traditional simulation modelling methods (e.g., DES, SD, ABS), with Big Data Analytics and artificial intelligence techniques (e.g., Genetic Algorithms, Artificial Neural Networks) in different ways to deal with the complexity of systems in Industry 4.0. Third, researchers are trying to add functionalities or develop new simulation software tools, such as in [41], which developed an integration mechanism to allow online communication between different software tools. Fourth, since Industry 4.0 is still in its infancy, some researchers are attempting to develop experiments (e.g., in learning factories) to show the potential of modelling and simulation to improve companies' operational performance and its different KPIs [42] to deploy Industry 4.0.

Since modelling and simulation in Industry 4.0 is an emerging field, there are several research gaps to be covered, such as the lack of general frameworks and efficient modelling and simulation methods and techniques, as well as software tools to support faster synchronization of the physical and virtual worlds, real-time data, real-time 3D simulation, and real-time decision making [23, 36]. There is also a need for semantic technologies to standardize systems' data format and communication protocols. In addition, for modelling and simulation in Industry 4.0 to gain more practical relevance, more controlled experiments and case studies are needed, together with empirical research to assess the impact of Industry 4.0 modelling and simulation in companies' operations performance and product life-cycle management (PLM), thus advancing initial efforts provided in [37].

The studies pointed out many research directions or opportunities: the building of new modelling and simulation frameworks to combine lean production tools, such as VSM, with Industry 4.0 technologies to assess the risks, costs, revenue potential and implementation barriers to deploy Industry 4.0, covering part of the gaps discussed in [3, 21, 38]; the development of practical cases by incorporating Big Data in the simulation models [19, 20]; and the development and implementation of more realistic Digital Twins (DT) throughout the supply chain and product life-cycle management (PLM) [23, 25, 27].

5.5 Conclusion

Managing complex systems is one of the eight priority areas for action set out in the final report from the German *Industrie 4.0* working group [4] and modelling and simulation is at its core as an enabling technology [5]. Some initial empirical research and patent analysis also suggest that modelling and simulation is one of the key technologies to produce innovations and deploy Industry 4.0 [37, 38, 43]. This paper therefore provides a systematic literature review and meta-analysis of 112 peer-reviewed publications in the field of modelling and simulation applied in the context of Industry 4.0, with a focus on manufacturing and logistics systems. The results suggest that the simulation-base approaches applied in the context of the Fourth Industrial Revolution are mostly agent-based or mixed methods, mainly combining established simulation paradigms (e.g., DES, SD, ABS), CAx, and Artificial Intelligence (AI) techniques, whilst identify Digital Twin (DT) as the most promising simulation technology.

Acknowledgements This research is supported by the Federal Institute of Education, Science and Technology of Sao Paulo (IFSP) in Brazil and by the Natural Sciences and Engineering Research Council (NSERC) of Canada [grant numbers RGPIN-2018-06680].

References

1. Liao, Y., Deschamps, F., Loures, E.F.R., et al.: Past, present and future of industry 4.0 - a systematic literature review and research agenda proposal. Int. J. Prod. Res. (2017). https://doi.org/10.1080/00207543.2017.1308576
2. Mazali, T.: From industry 4.0 to society 4.0, there and back. AI Soc. (2018). https://doi.org/10.1007/s00146-017-0792-6
3. Hofmann, E., Rüsch, M.: Industry 4.0 and the current status as well as future prospects on logistics. Comput. Ind. (2017). https://doi.org/10.1016/j.compind.2017.04.002
4. Kagermann, H., Helbig, J., Hellinger, A., et al.: Recommendations for implementing the strategic initiative INDUSTRIE 4.0: securing the future of German manufacturing industry; final report of the Industrie 4.0 Working Group. Forschungsunion, Frankfurt (2013)
5. Ghobakhloo, M.: The future of manufacturing industry: a strategic roadmap toward industry 4.0. J. Manuf. Technol. Manag. (2018). https://doi.org/10.1108/JMTM-02-2018-0057
6. Lichtblau, K., Stich, V., Bertenrath, R., et al.: Industrie 4.0-readiness. Aachen, Cologne (2015)
7. De Souza Dutra, M.D., Da Conceição Jr., G., De Paula Ferreira, W., et al.: A customized transition towards smart homes: a fast framework for economic analyses. Appl. Energy (2020). https://doi.org/10.1016/j.apenergy.2020.114549
8. Tao, F., Zhang, H., Liu, A., et al.: Digital twin in industry: state-of-the-art. IEEE Trans. Ind. Inform. (2019). https://doi.org/10.1109/TII.2018.2873186

9. Shafer, S.M., Smunt, T.L.: Empirical simulation studies in operations management: context, trends, and research opportunities. J. Oper. Manag. (2004). https://doi.org/10.1016/j.jom.2004.05.002
10. Jahangirian, M., Eldabi, T., Naseer, A., et al.: Simulation in manufacturing and business: a review. Eur. J. Oper. Res. (2010). https://doi.org/10.1016/j.ejor.2009.06.004
11. Negahban, A., Smith, J.S.: Simulation for manufacturing system design and operation: literature review and analysis. J. Manuf. Syst. (2014). https://doi.org/10.1016/j.jmsy.2013.12.007
12. Da Costa, L.S., Lucio, W.D.S., Da Silva, A.M., et al.: Discrete simulation applied to the production process of electronic components. Indep. J. Manag. Prod. (2017). https://doi.org/10.14807/ijmp.v8i5.603
13. Bertrand, J.W.M., Fransoo, J.C.: Operations management research methodologies using quantitative modeling. Int. J. Oper. Prod. Manag. (2002). https://doi.org/10.1108/01443570210414338
14. Rodič, B.: Industry 4.0 and the new simulation modelling paradigm. Organizacija (2017). https://doi.org/10.1515/orga-2017-0017
15. Vieira, A.A.C., Dias, L.M.S., Santos, M.Y.: Setting an industry 4.0 research and development agenda for simulation - a literature review. Int. J. Simul. Model. (2018). https://doi.org/10.2507/IJSIMM17(3)429
16. Moher, D., Liberati, A., Tetzlaff, J., et al.: Preferred reporting items for systematic reviews and meta-analyses: the PRISMA statement. Ann. Intern. Med. (2009). https://doi.org/10.7326/0003-4819-151-4-200908180-00135
17. Wohlin, C.: Guidelines for snowballing in systematic literature studies and a replication in software engineering. In: Proceedings of the 18th International Conference on Evaluation and Assessment in Software Engineering, London, 12–14 May 2014 (2014)
18. Derler, P., Lee, E.A., Vincentelli, A.S.: Modeling cyber-physical systems. Proc. IEEE (2012). https://doi.org/10.1109/JPROC.20
19. Wang, S., Wan, J., Zhang, D., et al.: Towards smart factory for industry 4.0: a self-organized multi-agent system with big data based feedback and coordination. Comput. Netw. (2016). https://doi.org/10.1016/j.comnet.2015.12.017
20. Rosen, R., Von Wichert, G., Lo, G., et al.: About the importance of autonomy and digital twins for the future of manufacturing. IFAC-PapersOnLine (2015). https://doi.org/10.1016/j.ifacol.2015.06.141
21. Hehenberger, P., Vogel-Heuser, B., Bradley, D., et al.: Design, modelling, simulation and integration of cyber physical systems: methods and applications. Comput. Ind. (2017). https://doi.org/10.14807/ijmp.v8i5.597
22. Mosterman, P.J., Zander, J.: Industry 4.0 as a cyber-physical system study. Softw. Syst. Model. (2016). https://doi.org/10.1007/s10270-015-0493-x
23. Tao, F., Cheng, J., Qi, Q., et al.: Digital twin-driven product design, manufacturing and service with big data. Int. J. Adv. Manuf. Technol. (2018). https://doi.org/10.1007/s00170-017-0233-1
24. Uhlemann, T.H.J., Lehmann, C., Steinhilper, R.: The digital twin: realizing the cyber-physical production system for industry 4.0. Procedia CIRP (2017). https://doi.org/10.1016/j.procir.2016.11.152
25. Schleich, B., Anwer, N., Mathieu, L., et al.: Shaping the digital twin for design and production engineering. CIRP Ann. (2017). https://doi.org/10.1016/j.procir.2016.11.152
26. Turner, C.J., Hutabarat, W., Oyekan, J., et al.: Discrete event simulation and virtual reality use in industry: new opportunities and future trends. IEEE Trans. Hum.-Mach. Syst. (2016). https://doi.org/10.1109/THMS.2016.2596099
27. Tao, F., Zhang, M.: Digital twin shop-floor: a new shop-floor paradigm towards smart manufacturing. IEEE Access (2017). https://doi.org/10.1109/ACCESS.2017.2756069
28. Fortino, G., Gravina, R., Russo, W., et al.: Modeling and simulating internet-of-things systems: a hybrid agent-oriented approach. Comput. Sci. Eng. (2017). https://doi.org/10.1109/MCSE.2017.3421541

29. Uhlemann, T.H.J., Schock, C., Lehmann, C., et al.: The digital twin: demonstrating the potential of real time data acquisition in production systems. Procedia Manuf. (2017). https://doi.org/10.1016/j.promfg.2017.04.043
30. Qi, Q., Tao, F.: Digital twin and big data towards smart manufacturing and industry 4.0: 360 degree comparison. IEEE Access (2018). https://doi.org/10.1109/ACCESS.2018.2793265
31. Penas, O., Plateaux, R., Patalano, S., et al.: Multi-scale approach from mechatronic to cyber-physical systems for the design of manufacturing systems. Comput. Ind. (2017). https://doi.org/10.1016/j.compind.2016.12.001
32. Weyer, S., Meyer, T., Ohmer, M., et al.: Future modeling and simulation of CPS-based factories: an example from the automotive industry. IFAC-PapersOnLine (2016). https://doi.org/10.1016/j.ifacol.2016.12.168
33. Van Eck, N.J., Waltman, L.: Visualizing bibliometric networks. In: Ding, Y., Rousseau, R., Wolfram, D. (eds.) Measuring Scholarly Impact, pp. 285–320. Springer, Switzerland (2014)
34. Tako, A.A., Robinson, S.: The application of discrete event simulation and system dynamics in the logistics and supply chain context. Decis. Support Syst. (2012). https://doi.org/10.1016/j.dss.2011.11.015
35. De Oliveira, C.A.S., De Paula Ferreira, W., Mendes, R.C., et al.: Metaheuristic ILS with path relinking for the number partitioning problem. Indep. J. Manag. Prod. (2017). https://doi.org/10.14807/ijmp.v8i5.597
36. Lee, J., Noh, S., Kim, H.J., et al.: Implementation of cyber-physical production systems for quality prediction and operation control in metal casting. Sensors (2018). https://doi.org/10.3390/s18051428
37. Jeong, J.Y., Han, Y., Kim, J., et al.: Empirical study of engineering modeling and simulation in manufacturing innovation to lead 4th industrial revolution. ICIC-ELB (2018). https://doi.org/10.24507/icicelb.09.05.421
38. Lugert, A., Batz, A., Winkler, H.: Empirical assessment of the future adequacy of value stream mapping in manufacturing industries. J. Manuf. Technol. Manag. (2018). https://doi.org/10.1108/JMTM-11-2017-0236
39. Scheidegger, A.P.G., Pereira, T.F., De Oliveira, M.L.M., et al.: An introductory guide for hybrid simulation modelers on the primary simulation methods in industrial engineering identified through a systematic review of the literature. Comput. Ind. Eng. (2018). https://doi.org/10.1016/j.cie.2018.07.046
40. Abar, S., Theodoropoulos, G.K., Lemarinier, P., et al.: Agent based modelling and simulation tools: a review of the state-of-art software. Comput. Sci. Rev. (2017). https://doi.org/10.1016/j.cosrev.2017.03.001
41. Laurindo, Q.M.G., Peixoto, T.A., De Assis Rangel, J.J.: Communication mechanism of the discrete event simulation and the mechanical project softwares for manufacturing systems. J. Comput. Des. Eng. (2018). https://doi.org/10.1016/j.jcde.2018.02.005
42. Kaganski, S., Majak, J., Karjust, K.: Fuzzy AHP as a tool for prioritization of key performance indicators. Procedia CIRP (2018). https://doi.org/10.1016/j.procir.2018.03.097
43. Han, Y., Jeong, J., Ko, M.H., et al.: Analysis of global competitiveness of engineering modeling and simulation technology for next-manufacturing innovation: using quantitative analysis of patents and papers. ICIC-ELB (2018). https://doi.org/10.24507/icicelb.09.04.339

Chapter 6
Homogenization Algorithm Based on Incremental L2-Discrepancy Filtering for Data-Driven Modelling

Zhuang Shao, Fengqi Si, Zhigao Xu, and Ding Guo

6.1 Introduction

The themes of Industry 4.0 (the fourth industrial revolution) distinguishing from the previous three are machine to machine communication (M2M), Internet of Things (IoT) [1, 2], cloud computing [3, 4] and artificial intelligence (AI) [5]. Behind these keywords, data is as important as blood for bones and muscles. At present, decentralized control systems and large databases are becoming a standard in enterprises, where terabytes of historical operating data in time series are stored and analyzed [6, 7]. A large number of applications in the era of Industry 4.0, such as equipment performance monitoring [8], predicting [9, 10], characteristic analysis [11], status warning [12] and fault diagnosis [13–15], have been developed based on the operation data.

However, due to the inflexible operation mode of the industrial system, there is a serious imbalance problem in the raw data. The data imbalance problem refers to the uneven distribution of samples in feature space [16, 17]. When the samples are dominated by a single pattern, the model is unreasonable and deviate from the real relationships among variables, even though the training performance index is good.

Z. Shao · F. Si (✉) · Z. Xu
Key Laboratory of Energy Thermal Conversion and Control of Ministry of Education, School of Energy and Environment, Southeast University, Nanjing, China
e-mail: fqsi@seu.edu.cn

Z. Shao
e-mail: zhuangshao@seu.edu.cn

Z. Xu
e-mail: zgxu@seu.edu.cn

D. Guo
Zhejiang Energy Group Research Institute, Hangzhou, China
e-mail: rorchi3223145@126.com

A. Dingli et al. (eds.), *Artificial Intelligence in Industry 4.0*,
Studies in Computational Intelligence 928,
https://doi.org/10.1007/978-3-030-61045-6_6

The data imbalance problem is harmful to most kinds of machine learning methods, which take the classification or regression error for all or batch of training samples as the loss function. The predict error of each sample is assigned the same weight. In the crowded region of feature space, the algorithm has less tolerance to the predict error, where even over-fitting occurs. As for the sparse region of feature space, the algorithm has more tolerance to the predict error, where even under-fitting occurs. In industrial systems, working conditions are generally determined by external factors, like production scheduling command and environment. For stable operation, most of the collected data is concentrated in a small region or in a small pattern, leading to serious data imbalance.

There are three methods to solve the data imbalance problem: tune the sample distribution [18], design new algorithms [19] and design new evaluation indicators [20]. Tuning the sample distribution is the most widespread one, where the over-sampling and under-sampling techniques are used. In this method, over-sampling is to increase minority samples to keep the balance between minority and majority. SMOTE (Synthetic Minority Over-sampling Technique) is a classic oversampling data processing method [20–22], which may lead to over-fitting [23]. Under-sampling is to remove majority samples, which has a risk of losing the characteristics of majority samples [24]. L2-discrepancy is an indicator for evaluating the uniformity of data distribution, proposed by Hickernell et al. [25, 26], and has a good application prospect in solving data imbalance problem. With the development of calculating power and algorithms, such as deep learning [27, 28], the sample size is also increasing [29]. The original L2-discrepancy has a large computational complexity when dealing with large size data sets, so that the computational time-consuming cannot meet the requirements of processing real-time data streams. This paper proposes an incremental L2-discrepancy algorithm, which reduces the time complexity of the original algorithm from $O(n^2)$ to $O(n)$. Furthermore, a novel under-sampling homogenization algorithm called incremental L2-discrepancy filtering (IL2DF) is proposed for mass data-driven modelling. Finally, an application case of modelling task in thermal power system is carried out, which proves the effectiveness of the proposed algorithm.

6.2 Methodology

Data distribution analysis a practical method to solve the data imbalance problem. L2-discrepancy evaluates the distribution of data from the perspective of sample distance. This section proposes the expression of L2-discrepancy, the derivation of incremental L2-discrepancy and the IL2DF.

6.2.1 L2-Discrepancy

Given a dataset with samples $x_i = [x_{i1}, x_{i2}, \ldots, x_{ip}]$, $i = 1, 2, \ldots, n$, normalized into [0, 1], where n is the number, and p is the number of features, L2-discrepancy is described in Eq. 6.1 [25, 26].

$$\lambda = \left[\left(\frac{13}{12} \right)^p - \frac{2^{1-p}}{n} \sum_{k=1}^{n} \prod_{i=1}^{p} \left(2 + \left| x_{ki} - \frac{1}{2} \right| - \left| x_{ki} - \frac{1}{2} \right|^2 \right) + \frac{1}{n^2} \sum_{k,l=1}^{n} \prod_{i=1}^{p} \left(1 + \frac{1}{2} \left| x_{ki} - \frac{1}{2} \right| + \frac{1}{2} \left| x_{li} - \frac{1}{2} \right| - \frac{1}{2} \left| x_{ki} - x_{li} \right| \right) \right]^{\frac{1}{2}} \tag{6.1}$$

where $x_{i,j}$ is the normalized sample. The L2-discrepancy $\lambda > 0$, and a smaller λ represents a more uniform distribution. The computational complexity of L2-discrepancy is $O(n^2)$ and the computational time is proportional to the square of the sample number n.

6.2.2 Incremental L2-Discrepancy

Incremental L2-discrepancy is to calculate L2-discrepancy of dataset X_{n+1} after adding a new sample of $x_{n+1} = [x_{n+1,1}, x_{n+1,2}, \ldots, x_{n+1,p}]$ to the original set $X_n = \{x_{ij} | i = 1, 2, \ldots, n; j = 1, 2, \ldots, p\}$, with a known L2-discrepancy of X_n. Mark the L2-discrepancy of the original set X_n as in Eq. 6.2.

$$\lambda_n^2 = \lambda_{na} + \lambda_{nb} + \lambda_{nc} \tag{6.2}$$

where

$$\lambda_{na} = \frac{13}{12}^p \tag{6.3}$$

$$\lambda_{nb} = -\frac{2^{1-p}}{n} \sum_{k=1}^{n} \prod_{i=1}^{p} \left(2 + \left| x_{ki} - \frac{1}{2} \right| - \left| x_{ki} - \frac{1}{2} \right| \right) \tag{6.4}$$

$$\lambda_{nc} = \frac{1}{n^2} \sum_{k,l=1}^{n} \prod_{i=1}^{p} \left(1 + \frac{1}{2} \left| x_{ki} - \frac{1}{2} \right| + \frac{1}{2} \left| x_{li} - \frac{1}{2} \right| - \frac{1}{2} \left| x_{ki} - x_{li} \right| \right) \tag{6.5}$$

Mark the L2-discrepancy of the new set X_{n+1} as in Eq. 6.6.

$$\lambda_{n+1}^2 = \lambda_{n+1,a} + \lambda_{n+1,b} + \lambda_{n+1,c} \tag{6.6}$$

where

$$\lambda_{n+1,a} = \frac{13^p}{12} \tag{6.7}$$

$$\lambda_{n+1,b} = -\frac{2^{1-p}}{n+1}\sum_{k=1}^{n+1}\prod_{i=1}^{p}\left(2+\left|x_{ki}-\frac{1}{2}\right|-\left|x_{ki}-\frac{1}{2}\right|\right) \tag{6.8}$$

$$\lambda_{n+1,c} = \frac{1}{n+1^2}\sum_{k,l=1}^{n+1}\prod_{i=1}^{p}\left(1+\frac{1}{2}\left|x_{ki}-\frac{1}{2}\right|+\frac{1}{2}\left|x_{li}-\frac{1}{2}\right|-\frac{1}{2}\left|x_{ki}-x_{li}\right|\right) \tag{6.9}$$

Solve λ_{n+1} as in Eq. 6.10:

$$\begin{aligned}\lambda_{n+1} &= [\lambda_n^2 - (\lambda_{na}+\lambda_{nb}+\lambda_{nc}) + (\lambda_{n+1,a}+\lambda_{n+1,b}+\lambda_{n+1,c})]^{\frac{1}{2}} \\ &= [\lambda_n^2 + (\lambda_{n+1,a}-\lambda_{na}) + (\lambda_{n+1,b}-\lambda_{nb}) + (\lambda_{n+1,c}-\lambda_{nc})]^{\frac{1}{2}}\end{aligned} \tag{6.10}$$

where

$$\lambda_{n+1,a} - \lambda_{na} = 0 \tag{6.11}$$

Transform the form of $\lambda_{n+1,b} - \lambda_{nb}$ as in Eq. 6.12.

$$\begin{aligned}\lambda_{n+1,b} - \lambda_{nb} &= -\frac{2^{1-p}}{n+1}\sum_{k=1}^{n+1}\prod_{i=1}^{p}\left(2+\left|x_{ki}-\frac{1}{2}\right|-\left|x_{ki}-\frac{1}{2}\right|^2\right) \\ &\quad+\frac{2^{1-p}}{n}\sum_{k=1}^{n}\prod_{i=1}^{p}\left(2+\left|x_{ki}-\frac{1}{2}\right|-\left|x_{ki}-\frac{1}{2}\right|^2\right) \\ &= -\frac{2^{1-p}}{n+1}[\sum_{k=1}^{n}\prod_{i=1}^{p}\left(2+\left|x_{ki}-\frac{1}{2}\right|-\left|x_{ki}-\frac{1}{2}\right|^2\right) \\ &\quad+\prod_{i=1}^{p}\left(2+\left|x_{n+1,i}-\frac{1}{2}\right|-\left|x_{n+1,i}-\frac{1}{2}\right|^2\right)] \\ &\quad+\frac{2^{1-p}}{n}\sum_{k=1}^{n}\prod_{i=1}^{p}\left(2+\left|x_{ki}-\frac{1}{2}\right|-\left|x_{ki}-\frac{1}{2}\right|^2\right) \\ &= \sum_{k=1}^{n}\prod_{i=1}^{p}\left(2+\left|x_{ki}-\frac{1}{2}\right|-\left|x_{ki}-\frac{1}{2}\right|^2\right)\left(-\frac{2^{1-p}}{n+1}+\frac{2^{1-p}}{n}\right) \\ &\quad-\frac{2^{1-p}}{n+1}\prod_{i=1}^{p}\left(2+\left|x_{n+1,i}-\frac{1}{2}\right|-\left|x_{n+1,i}-\frac{1}{2}\right|^2\right) \\ &= -\frac{1}{n+1}\lambda_{nb} - \frac{2^{1-p}}{n+1}\prod_{i=1}^{p}\left(2+\left|x_{n+1,i}-\frac{1}{2}\right|-\left|x_{n+1,i}-\frac{1}{2}\right|^2\right)\end{aligned} \tag{6.12}$$

Transform the form of $\lambda_{n+1,c} - \lambda_{nc}$ as in Eq. 6.13.

$$
\begin{aligned}
&\lambda_{n+1,c} - \lambda_{nc} \\
&= \frac{1}{(n+1)^2} \sum_{k,l=1}^{n+1} \prod_{i=1}^{p} \left(1 + \frac{1}{2}\left|x_{ki} - \frac{1}{2}\right| + \frac{1}{2}\left|x_{li} - \frac{1}{2}\right| - \frac{1}{2}\left|x_{ki} - x_{li}\right|\right) \\
&\quad - \frac{1}{n^2} \sum_{k,l=1}^{n} \prod_{i=1}^{p} \left(1 + \frac{1}{2}\left|x_{ki} - \frac{1}{2}\right| + \frac{1}{2}\left|x_{li} - \frac{1}{2}\right| - \frac{1}{2}\left|x_{ki} - x_{li}\right|\right) \\
&= \frac{1}{(n+1)^2} [\sum_{k,l=1}^{n} \prod_{i=1}^{p} \left(1 + \frac{1}{2}\left|x_{ki} - \frac{1}{2}\right| + \frac{1}{2}\left|x_{li} - \frac{1}{2}\right| - \frac{1}{2}\left|x_{ki} - x_{li}\right|\right) \\
&\quad + \sum_{l=1}^{n} \prod_{i=1}^{p} \left(1 + \frac{1}{2}\left|x_{n+1,i} - \frac{1}{2}\right| + \frac{1}{2}\left|x_{li} - \frac{1}{2} - \frac{1}{2}\left|x_{n+1,i} - x_{li}\right|\right|\right) \\
&\quad + \sum_{l=1}^{n} \prod_{i=1}^{p} \left(1 + \frac{1}{2}\left|x_{ki} - \frac{1}{2}\right| + \frac{1}{2}\left|x_{n+1,i} - \frac{1}{2} - \frac{1}{2}\left|x_{ki} - x_{n+1,i}\right|\right|\right) \\
&\quad + \prod_{i=1}^{p} \left(1 + \frac{1}{2}\left|x_{n+1,i} - \frac{1}{2}\right| + \frac{1}{2}\left|x_{n+1,i} - \frac{1}{2} - \frac{1}{2}\left|x_{n+1,i} - x_{n+1,i}\right|\right|\right) \\
&\quad - \frac{1}{n^2} \sum_{k,l=1}^{n} \prod_{i=1}^{p} \left(1 + \frac{1}{2}\left|x_{ki} - \frac{1}{2}\right| + \frac{1}{2}\left|x_{li} - \frac{1}{2} - \frac{1}{2}\left|x_{ki} - x_{li}\right|\right|\right) \\
&= \sum_{k,l=1}^{n} \prod_{i=1}^{p} \left(1 + \frac{1}{2}\left|x_{ki} - \frac{1}{2}\right| + \frac{1}{2}\left|x_{li} - \frac{1}{2}\right| - \frac{1}{2}\left|x_{ki} - x_{li}\right|\right) \left(\frac{1}{(n+1)^2} - \frac{1}{n^2}\right) \\
&\quad + \frac{1}{(n+1)^2} \left[2 \sum_{k=1}^{n} \prod_{i=1}^{p} \left(1 + \frac{1}{2}\left|x_{ki} - \frac{1}{2}\right| + \frac{1}{2}\left|x_{n+1,i} - \frac{1}{2}\right| - \frac{1}{2}\left|x_{ki} - x_{n+1,i}\right|\right)\right] \\
&\quad + \frac{1}{(n+1)^2} \prod_{i=1}^{p} \left(1 + \left|x_{n+1,i} - \frac{1}{2}\right|\right) \\
&= -\frac{2n+1}{(n+1)^2} \lambda_{nc} \\
&\quad + \frac{2}{(n+1)^2} \sum_{k=1}^{n} \prod_{i=1}^{p} \left(1 + \frac{1}{2}\left|x_{ki} - \frac{1}{2}\right| + \frac{1}{2}\left|x_{n+1,i} - \frac{1}{2}\right| - \frac{1}{2}\left|x_{ki} - x_{n+1,i}\right|\right) \\
&\quad + \frac{1}{(n+1)^2} \prod_{i=1}^{p} \left(1 + \left|x_{n+1,i} - \frac{1}{2}\right|\right)
\end{aligned}
\tag{6.13}
$$

Then, λ_{n+1} is calculated in Eq. 6.14.

$$\begin{aligned}\lambda_{n+1} = [\lambda_n^2 &- \frac{1}{n+1}\lambda_{nb} - \frac{2^{1-p}}{n+1}\prod_{i=1}^{p}\left(2 + \left|x_{n+1,i} - \frac{1}{2}\right| - \left|x_{n+1,i} - \frac{1}{2}\right|^2\right) - \frac{2n+1}{(n+1)^2}\lambda_{nc} \\ &+ \frac{2}{(n+1)^2}\sum_{k=1}^{n}\prod_{i=1}^{p}\left(1 + \frac{1}{2}\left|x_{ki} - \frac{1}{2}\right| + \frac{1}{2}\left|x_{n+1,i} - \frac{1}{2}\right| - \frac{1}{2}\left|x_{ki} - x_{n+1,i}\right|\right) \\ &+ \frac{1}{(n+1)^2}\prod_{i=1}^{p}\left(1 + \left|x_{n+1,i} - \frac{1}{2}\right|\right)]^{\frac{1}{2}}\end{aligned} \tag{6.14}$$

where $\lambda_{n+1,a} = \lambda_{na}$, $\lambda_{n+1,b}$ equals the sum of λ_{nb} and items 2, 3 from bracket of Eq. 6.14. $\lambda_{n+1,c}$ equals the sum of λ_{nc} and items 4, 5, 6 from bracket of Eq. 6.14.

Using a similar approach, one can derivative the λ_{n-1} of dataset X_{n-1} after removing one existing sample of $x_n = [x_{n1}, x_{n2}, \ldots, x_{np}]$ from the original set $X_n = x_{ij}|i = 1, 2, \ldots, n; j = 1, 2, \ldots, p$, with a known L2-discrepancy of X_n, which is shown in Eq. 6.15.

$$\begin{aligned}\lambda_{n-1} = [\lambda_n^2 &- \frac{1}{n-1}\lambda_{nb} - \frac{2^{1-p}}{n-1}\prod_{i=1}^{p}\left(2 + \left|x_{ni} - \frac{1}{2}\right| - \left|x_{ni} - \frac{1}{2}\right|^2\right) - \frac{2n-1}{(n-1)^2}\lambda_{nc} \\ &- \frac{2}{(n-1)^2}\sum_{k=1}^{n}\prod_{i=1}^{p}\left(1 + \frac{1}{2}\left|x_{ki} - \frac{1}{2}\right| + \frac{1}{2}\left|x_{n,i} - \frac{1}{2}\right| - \frac{1}{2}\left|x_{ki} - x_{ni}\right|\right) \\ &+ \frac{1}{(n-1)^2}\prod_{i=1}^{p}\left(1 + \left|x_{ni} - \frac{1}{2}\right|\right)]^{\frac{1}{2}}\end{aligned} \tag{6.15}$$

where $\lambda_{n-1,a} = \lambda_{na}$, $\lambda_{n-1,b}$ equals the sum of λ_{nb} and items 2, 3 from bracket of Eq. 6.15. $\lambda_{n-1,c}$ equals the sum of λ_{nc} and items 4, 5, 6 from bracket of Eq. 6.15. It can be seen that the computational complexity of incremental L2-discrepancy is $O(n)$ and the computational time is proportional to the sample number n.

6.2.3 Incremental L2-Discrepancy Filtering

Homogenization algorithm can be designed based on incremental L2-discrepancy, which is proposed in Algorithm 1.

In addition to the random remove method, one can also design an algorithm to find the sample, which can reduce the most L2-discrepancy value, to be removed. However, it will increase the computational complexity of the framework. With IL2DF, it is possible to evaluate the change of L2-discrepancy after adding or removing a sample with little calculation cost, and decide whether to update the sample set.

Algorithm 1 Incremental L2-discrepancy filtering (IL2DF)

initialization: dataset D=null, m=0, λ=0, λ_a=0, λ_b=0, λ_c=0
repeat
  input a new sample x_{+1}, $m \leftarrow m + 1$
  if $m \leq m_{max}$ **then**
    D.add(x_{+1})
    if m==m_{max} **then**
      calculate λ,λ_a,λ_b,λ_c according to Eqs. 6.1, 6.2, 6.3, 6.4, 6.5
    end if
  else
    calculate λ_{+1}, $\lambda_{+1,a}$, $\lambda_{+1,b}$, $\lambda_{+1,c}$ according to Eq. 6.14 and x_{+1}, λ, λ_a, λ_b, λ_c
    if $\lambda_{+1} < \lambda$ **then**
      D.add(x_{+1})
      $\lambda \leftarrow \lambda_{+1}$,$\lambda_a \leftarrow \lambda_{+1,a}$,$\lambda_b \leftarrow \lambda_{+1,b}$,$\lambda_c \leftarrow \lambda_{+1,c}$
      ind = random integer from $[1, m_{max}]$
      calculate λ_{-1},$\lambda_{-1,a}$,$\lambda_{-1,b}$,$\lambda_{-1,c}$ according to Eq. 6.15 and x_{ind}, λ,λ_a,λ_b,λ_c
      $\lambda \leftarrow \lambda_{-1}$,$\lambda_a \leftarrow \lambda_{-1,a}$,$\lambda_b \leftarrow \lambda_{-1,b}$,$\lambda_c$
      $\lambda_{-1,c}$
      D.remove(x_{ind})
    end if
  end if
until finish

6.3 Cases

In order to prove the advantages of IL2DF, this paper designs three cases. Computational complexity analysis, standard distribution case and application case.

6.3.1 Computational Complexity Analysis

This section compares the computational time consuming about L2-discrepancy and incremental L2-discrepancy with the increasing number of samples in the dataset with 2 features.

From Fig. 6.1, with the increase of the number of samples in the data set, the original L2-discrepancy algorithm takes much longer than the incremental L2-discrepancy algorithm. When the number of samples reaches 10,000, the time consuming of original L2-discrepancy is more than 1 s, and the incremental L2-discrepancy is less than 10^{-3} s. When the number of samples reaches 30,000, the time consuming of original L2-discrepancy is more than 10 s, and the incremental L2-discrepancy is around 10^{-3} s. Data sets with more than 10,000 samples are common, but 10 s of computation time is unbearable facing the real-time data streams.

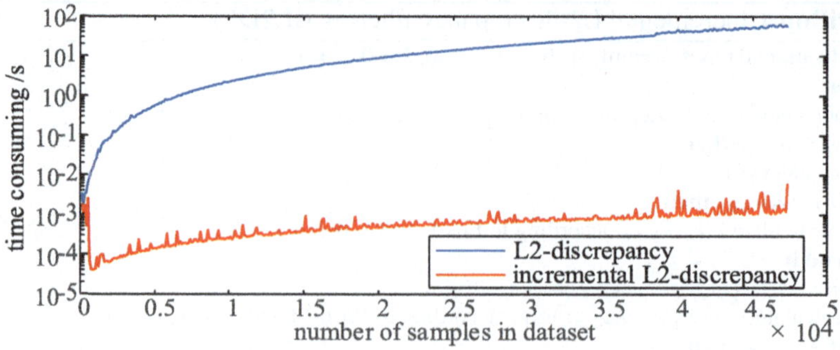

Fig. 6.1 Time-consuming of L2-discrepancy and incremental L2-discrepancy for growing dataset (*Source* Author)

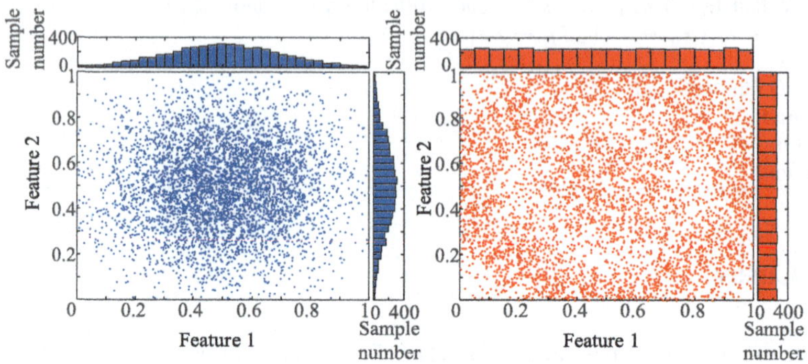

Fig. 6.2 Samples from a Two-dimensional normal distribution (blue, L2-discrepancy = 0.157) with the variance of 0.2, mean value of 0.5 and the amount of samples (red, L2-discrepancy = 0.005) filtered by IL2DF from the same distribution (*Source* Author)

6.3.2 Standard Distribution Case

This case demonstrates the distribution balance effect for IL2DF, facing an imbalanced data source.

It can be seen from Fig. 6.2 that the data distribution of data source from a normal distribution, after the processing of IL2DF, becomes more uniform. Its L2-discrepancy is reduced from 0.157 to 0.005. Data imbalance problem will no longer appear in modelling with these sampled data.

6.3.3 Application Case

In the industrial process, the distribution of system operating points is nonuniform, which limits the application of data-driven modelling. For example, in the perfor-

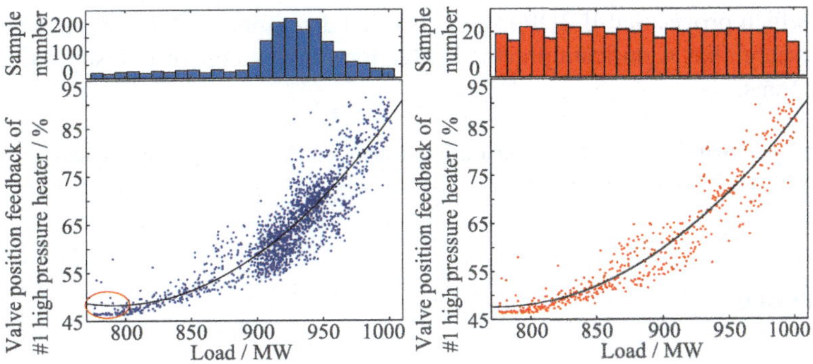

Fig. 6.3 Least-squares regression using raw data (blue) and samples (red) by IL2DF (*Source* Author)

mance analysis of thermal power units, the load is often selected as the main feature. However, the generator active power is controlled by AGC (Automatic Generation Control) system following the command from the power grid, which is determined by the power dispatch rule and electricity consumption of residents and factories. Consequently, modelling with raw data with externally determined distribution is very easy to cause data imbalance problem and model collapse. This case demonstrates the effect of IL2DF facing imbalance raw data (Fig. 6.3).

In the analysis of leakage failure of high-pressure heaters in power plants, the correlation modelling of drain water control valve and load is an important task [30, 31]. Generally, from the middle to high load range, the valve position and load have a quadratic relationship. Since the data is concentrated in [900, 950], the data in other interval is the minority, and the least-squares regression curve obtained is abnormal. Below 800 MW, the higher the load, the smaller the valve position, which is inconsistent with experience. In the load space, the L2-discrepancy of data set is 0.190. After the processing of IL2DF with the raw data, the data distribution of the load space becomes uniform. The L2-discrepancy of data set is 0.002, and the least-squares regression curve obtained is reasonable.

6.4 Conclusion

Data imbalance is a common problem for data-driven applications in modern factories for the lacking of diversity in operating mode, which results in unpredictable security risks. This paper introduces a distribution uniformity index L2-discrepancy, and develops its incremental form, which reduces the time complexity of the original algorithm from $O(n^2)$ to $O(n)$. Homogenization algorithm based on L2-discrepancy called incremental L2-discrepancy filtering (IL2DF) is proposed. The performance of the proposed algorithm is evaluated in computational complexity analysis, standard distribution case and application case about the correlation modelling in power plant, which proves that IL2DF is able to avoid date imbalance problem efficiently. This research has significant practical value for a wide range of Industry 4.0 related applications.

Acknowledgements This work was financially supported by Zhejiang Major Science and Technology Project (No. 2017C01082).

References

1. Mattern, F., Flörkemeier, C.: Vom Internet der Computer zum Internet der Dinge. Informatik-Spektrum **33**(2), 107–121 (2010). https://doi.org/10.1007/s00287-010-0417-7
2. Raji, R.S.: Smart networks for control. IEEE Spectr. **31**(6), 49–55 (1994). https://doi.org/10.1109/6.284793
3. Jasperneite, J.: Was hinter Begriffen wie Industrie 4.0 steckt. Comput. Autom. **19** (2012)
4. Kagermann, H., Helbig, J., Hellinger, A., Wahlster, W.: Recommendations for implementing the strategic initiative industrie 4.0: securing the future of German manufacturing industry; final report of the Industrie 4.0 Working Group. Forschungsunion (2013)
5. Tesch da Silva, F.S., da Costa, C.A., Paredes Crovato, C.D., da Rosa Righi, R.: Looking at energy through the lens of Industry 4.0: A systematic literature review of concerns and challenges. Comput. Ind. Eng. **143**, 106426 (2020). https://doi.org/10.1016/j.cie.2020.106426
6. Lee, J., Ardakani, H.D., Yang, S., Bagheri, B.: Industrial big data analytics and cyber-physical systems for future maintenance and service innovation. Procedia CIRP **38**, 3–7 (2015). https://doi.org/10.1016/j.procir.2015.08.026
7. ur Rehman, M.H., Yaqoob, I., Salah, K., Imran, M., Jayaraman, P.P., Perera, C.: The role of big data analytics in industrial internet of things. Futur. Gener. Comput. Syst. **99**, 247–259 (2019). https://doi.org/10.1016/j.future.2019.04.020
8. Mörth, O., Emmanouilidis, C., Hafner, N., Schadler, M.: Cyber-physical systems for performance monitoring in production intralogistics. Comput. Ind. Eng. **142**, 106333 (2020). https://doi.org/10.1016/j.cie.2020.106333
9. Kuo, C.-J., Ting, K.-C., Chen, Y.-C., Yang, D.-L., Chen, H.-M.: Automatic machine status prediction in the era of industry 4.0: case study of machines in a spring factory. J. Syst. Architect. **81**, 44–53 (2017). https://doi.org/10.1016/j.sysarc.2017.10.007
10. Yao, L., Ge, Z.: Big data quality prediction in the process industry: a distributed parallel modeling framework. J. Process Control **68**, 1–13 (2018). https://doi.org/10.1016/j.jprocont.2018.04.004
11. Moore, K.J.: Characteristic nonlinear system identification: a data-driven approach for local nonlinear attachments. Mech. Syst. Signal Process. **131**, 335–347 (2019). https://doi.org/10.1016/j.ymssp.2019.05.066
12. Vafaei, N., Ribeiro, R.A., Camarinha-Matos, L.M.: Fuzzy early warning systems for condition based maintenance. Comput. Ind. Eng. **128**, 736–746 (2019). https://doi.org/10.1016/j.cie.2018.12.056
13. Li, C., Zhang, S., Qin, Y., Estupinan, E.: A systematic review of deep transfer learning for machinery fault diagnosis. Neurocomputing **407**, 121–135 (2020). https://doi.org/10.1016/j.neucom.2020.04.045
14. Wang, Y., Wu, D., Yuan, X.: LDA-based deep transfer learning for fault diagnosis in industrial chemical processes. Comput. Chem. Eng. **140**, 106964 (2020). https://doi.org/10.1016/j.compchemeng.2020.106964
15. Hoang, D.-T., Kang, H.-J.: A survey on Deep Learning based bearing fault diagnosis. Neurocomputing **335**, 327–335 (2019). https://doi.org/10.1016/j.neucom.2018.06.078

16. Chawla, N.V.: Data mining for imbalanced datasets: an overview. Data Mining and Knowledge Discovery Handbook (pp. 875–886). Springer US (2009). https://doi.org/10.1007/978-0-387-09823-4_45
17. He, Haibo, Garcia, E.A.: Learning from imbalanced data. IEEE Trans. Knowl. Data Eng. **21**(9), 1263–1284 (2009). https://doi.org/10.1109/tkde.2008.239
18. Batista, G.E.A.P.A., Prati, R.C., Monard, M.C.: A study of the behavior of several methods for balancing machine learning training data. ACM SIGKDD Explor. Newsl. **6**(1), 20–29 (2004). https://doi.org/10.1145/1007730.1007735
19. López, V., Fernández, A., García, S., Palade, V., Herrera, F.: An insight into classification with imbalanced data: empirical results and current trends on using data intrinsic characteristics. Inf. Sci. **250**, 113–141 (2013). https://doi.org/10.1016/j.ins.2013.07.007
20. Chawla, N.V., Bowyer, K.W., Hall, L.O., Kegelmeyer, W.P.: SMOTE: synthetic minority over-sampling technique. J. Artif. Intell. Res. **16**, 321–357 (2002). https://doi.org/10.1613/jair.953
21. Deng, X., Zhong, W., Ren, J., Zeng, D., Zhang, H.: An imbalanced data classification method based on automatic clustering under-sampling. In: IEEE 35th International Performance Computing and Communications Conference (IPCCC), pp. 1–8. Las Vegas, NV (2016). https://doi.org/10.1109/PCCC.2016.7820640
22. Fernandez, A., Garcia, S., Herrera, F., Chawla, N.V.: SMOTE for learning from imbalanced data: progress and challenges, marking the 15-year anniversary. J. Artif. Intell. Res. **61**, 863–905 (2018). https://doi.org/10.1613/jair.1.11192
23. Sun, Z., Song, Q., Zhu, X., Sun, H., Xu, B., Zhou, Y.: A novel ensemble method for classifying imbalanced data. Pattern Recogn. **48**(5), 1623–1637 (2015). https://doi.org/10.1016/j.patcog.2014.11.014
24. Zhu, M., Xu, C., Wu, Y.-F.B.: IFME. In: Proceedings of the 13th ACM/IEEE-CS Joint Conference on Digital Libraries - JCDL'13. Presented at the the 13th ACM/IEEE-CS joint conference (2013). https://doi.org/10.1145/2467696.2467736
25. Hickernell, F.J.: A generalized discrepancy and quadrature error bound. Math. Comput. Am. Math. Soc. **67**(221), 299–322 (1998). https://doi.org/10.1090/s0025-5718-98-00894-1
26. Lv, Y., Romero, C.E., Yang, T., Fang, F., Liu, J.: Typical condition library construction for the development of data-driven models in power plants. Appl. Therm. Eng. **143**, 160–171 (2018). https://doi.org/10.1016/j.applthermaleng.2018.07.083
27. Sun, C., Shrivastava, A., Singh, S., Gupta, A.: Revisiting unreasonable effectiveness of data in deep learning era. In Proceedings of the IEEE International Conference on Computer Vision, pp. 843–852 (2017)
28. Zhang, Q., Yang, L.T., Chen, Z., Li, P.: A survey on deep learning for big data. Inf. Fusion **42**, 146–157 (2018). https://doi.org/10.1016/j.inffus.2017.10.006
29. Najafabadi, M.M., Villanustre, F., Khoshgoftaar, T.M., Seliya, N., Wald, R., Muharemagic, E.: Deep learning applications and challenges in big data analytics. J. Big Data **2**(1) (2015). https://doi.org/10.1186/s40537-014-0007-7
30. Zheng, M., Qian, H., Lin, S., Xiao, B., Chu, X.: Research on expert knowledge base of intelligent diagnosis based on tubing leakage of high-pressure heater in nuclear power plant. Advanced Computational Methods in Energy, Power, Electric Vehicles, and Their Integration, pp. 185–194. Springer, Singapore (2017)
31. Heo, G., Lee, S.K.: Internal leakage detection for feedwater heaters in power plants using neural networks. Expert Syst. Appl. **39**(5), 5078–5086 (2012). https://doi.org/10.1016/j.eswa.2011.11.031

Chapter 7
Linking Industry 4.0, Learning Factory and Simulation: Testbeds and Proof-of-Concept Experiments

William de Paula Ferreira, Arun Palaniappan, Fabiano Armellini, Luis Antonio de Santa-Eulalia, Elaine Mosconi, and Guillaume Marion

Abstract Learning factories have an important role in the development of Industry 4.0, providing a rich environment where researchers and companies can collaborate and test (in multiple scenarios) the application of cutting-edge technologies (e.g. the internet of things, big data, collaborative robots, additive manufacturing, modeling and simulation). The purpose of this paper is to identify and describe testbeds and proof-of-concept experiments, mainly related to modeling and simulation, developed in learning factories to support Industry 4.0 deployment. It also aims to present the lab shared by Produtique Québec and the Center of Excellence in Innovative Manufacturing Enterprise Management (CEGEMI), located at Sherbrooke, Canada, identifying its potential to conduct simulation-based research and to promote the digitalization of manufacturing companies. The literature review and case presentation show the relevance of leaning factories to develop applied research and to deploy Industry 4.0, especially in small and medium-sized enterprises (SME) that incrementally tend to move towards digitalization. Moreover, it describes how learning

W. de Paula Ferreira (✉) · F. Armellini
Polytechnique Montreal, Montreal, QC, Canada
e-mail: william.ferreira@polymtl.ca

F. Armellini
e-mail: fabiano.armellini@polymtl.ca

A. Palaniappan
NITT, Tiruchirappalli 620015, Tamil Nadu, India
e-mail: arunpalaniappan98@gmail.com

L. A. de Santa-Eulalia · E. Mosconi
Université de Sherbrooke, Sherbrooke, QC, Canada
e-mail: l.santa-eulalia@usherbrooke.ca

E. Mosconi
e-mail: elaine.Mosconi@usherbrooke.ca

G. Marion
CEGEMI, Sherbrooke, QC, Canada
e-mail: guillaume.marion@cegemi.org

A. Dingli et al. (eds.), *Artificial Intelligence in Industry 4.0*,
Studies in Computational Intelligence 928,
https://doi.org/10.1007/978-3-030-61045-6_7

factories can incorporate and test a wide range of Industry 4.0' technologies through testbeds and proof-of-concept experiments, supporting experimental validation of different artifacts, such as modeling and simulation frameworks.

7.1 Introduction

On the one hand, Industry 4.0 (I4.0) is one of the main strategies to improve the competitiveness of the manufacturing sector over the next years [1, 2]. It will lead to the emergence of dynamic, real-time optimized, self-organizing systems, making the creation of new business models based on mass customization possible [1]. However, I4.0 (or the Fourth Industrial Revolution) is still in its infancy [2]. Moreover, most legacy systems are not Industry 4.0-ready [3, 4], and companies tend to deploy I4.0 incrementally, especially small and medium-sized enterprises (SME) [5]. The development of empirical and applied research on I4.0 with real cases is still limited, which is an issue to advance the research field further. On the other hand, learning factories can play an important role in the development of I4.0, since they provide a rich environment where researchers and companies can collaborate and test (in multiple scenarios) the application of cutting-edge technologies (e.g. the internet of things, big data, collaborative robots, additive manufacturing, modeling and simulation). The concept of a learning factory emerged in 1994 in the US but has evolved over the years [6]. There are different morphologies and typologies of learning factories in the literature. The multidimensional description model proposed by the International Academy for Production Engineering Collaborative Working Group on Learning Factories involves 59 characteristics, grouped in 7 categories (i.e. operating model, purpose/target, process, setting, product, didactics, and metrics) [7]. Overall, a learning factory aims to provide a close-to-industry real environment for education and training as well as supporting collaboration among different stakeholders, research, development, technological transference, and innovation [6].

Learning factories are mainly present in universities, industries, consulting companies, and vocational schools [8]. Similar to I4.0, a learning factory can be considered as a socio-technical system [2, 8]. Concerning the technological infrastructure, a learning factory can assume different levels of configuration (i.e., station, cell, system, segment, factory, network) and adopt different products (i.e. real products, didactically prepared products, fantasy products, self-developed products) [8]. However, it must be as close as possible to reality and allow experiential learning approaches [6]. Learning factories may have different purposes, such as production process improvement, energy and resource efficiency, sustainability, logistics optimization, business administration, emergency processes, demonstrations, and more [6, 7]. There are also different learning factory environments, i.e. physical, digital, virtual, or hybrid environments, where the latest is more significant in the context of I4.0 [6]. Regarding the scope of this research, it is limited to I4.0 learning factories with a focus on research, technology transfer, and innovation.

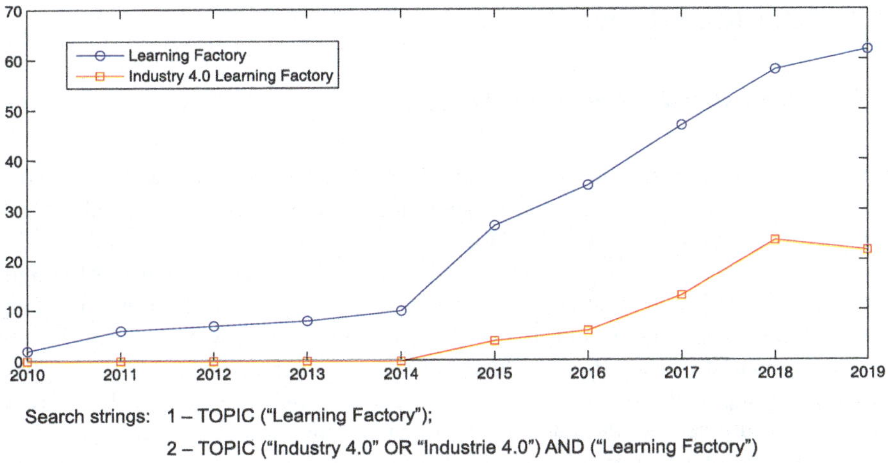

Fig. 7.1 Publications over the last 10 years (*Source* Authors)

According to Abel et al. [6, p. 809], "learning factories are more and more used as test areas for research," specially I4.0 learning factories. However, existing literature reviews on I4.0 learning factories focus on theoretical studies or descriptive cases of underdevelopment learning factories rather than how I4.0 learning factories is applied to conducting applied research [6, 7]. Hence, the purpose of this paper is twofold: first, based on a literature review, identify and describe testbeds and proof-of-concept experiments, mainly related to modeling and simulation, developed in learning factories to support I4.0 deployment; second, to present the lab shared by Produtique Québec and its Center of Excellence in Innovative Manufacturing Enterprise Management (CEGEMI) in Sherbrooke, Canada, identifying its potential to conduct simulation-based research and to promote the digitalization of manufacturing companies. For the literature review, the systematic guide to literature review described in Okoli and Schabram [9] was adopted. The data were collected through electronic data sources (Web of Science Core Collection and Scopus) using two search strings (see Fig. 7.1 and Table 7.1) to analyze the development of I4.0 learning factories and to identify existing I4.0 testbeds and proof-of-concept experiments documented in the scientific literature. As for the case presentation (or lab description), we conducted site visits and semi-structured interviews with lab coordinators to understand how it can support applied research and companies' transition to I4.0.

The remainder of the paper is organized as follows: Sect. 7.2 introduces Industry 4.0 learning factories and describes key Industry 4.0 testbeds and proof-of-concept experiments. Section 7.3 presents the lab description. Section 7.4 discusses the potential use of the lab as a research setting and drawn the conclusions.

7.2 Literature Review

7.2.1 Industry 4.0 Learning Factories

There are over 12 design principles and 14 enabling technologies within the context of I4.0 [10, 11], covering different socio-technical dimensions (i.e., technology, organization, and personnel) [12]. However, conforming to Elbestawi et al. [13], only a few learning factories cover a wide range of I4.0 components. Although, along with I4.0 [2], the number of publications on I4.0 learning factories is growing over time (see Fig. 7.1). Similar to companies, most existing learning factories are not yet Industry 4.0-ready [6, 14]. Hence, there is a branch in the scientific literature seeking to support the transition of existing learning factories to I4.0. In Karre et al. [14], the authors describe a detailed roadmap to transform an existing Lean Manufacturing learning factory into an I4.0 learning factory. A transformation model based on three pillars (i.e., didactic, integration, and organization) is also proposed in Baena et al. [15].

The primary purpose of a learning factory is to provide a close-to-industry environment to improve knowledge transfer [8]. However, existing I4.0 learning factories are mainly directed to research and technological transfer [6, 16, 17]. In Elbestawi et al. [13], the authors describe an I4.0 learning factory at McMaster University, Canada. A modular and reconfigurable learning factory at Winston University, Canada, is reported in [18]. Other examples of I4.0 learning factories include the AutoFab at the Darmstadt University of Applied Science, Germany; the Smart Production Laboratory at Aalborg University, Denmark [16]; and the Industry 4.0 Pilot Factory in Vienna University of Technology, Austria [19].

7.2.2 Industry 4.0 Testbeds and Proof-of-Concepts

According to Tisch and Metternich [20, p. 92], "learning factories environments are often used as validation settings." There is an increasing number of scientific articles in the field of I4.0, including a testbed or a proof-of-concept experiment, mostly developed in learning factories (see Table 7.1). Testbeds are experimental platforms that provide means for testing different artifacts (e.g. models, methods) and new technologies. They are "composite abstraction of systems used to study system components and interactions to gain further insight into the essence of the real system" [21, p. 33]. While proof-of-concept "is framed in terms of a particular kind of research that provides justification in practice of the potential transportability of knowledge acquired through the experimental test case" [22, p. 737]. Further analysis of the term can be found in [23]. As summarized in Table 7.1, a list of key research papers on I4.0 applying or describing the development of a testbed or proof-of-concept experiments is analyzed in the next paragraphs.

Table 7.1 Industry 4.0 testbeds and proof-of-concept experiments

Reference	Approach	Technologies	Focus/Application
Zhou et al. [24]	Testbed	SM	A general framework for knowledge-driven digital twin of a manufacturing work cell
Delbrugger et al. [25]	Virtual testbed	SM	Intelligent manufacturing. Multi-level model for optimization of production systems
Tzimas et al. [26]	Proof-of-concept	AR	Machine tool setup instructions
Kim et al. [27]	Testbed	SM	Reconfigurable manufacturing systems
Priggemeyer et al. [28]	Testbed	SM	Simulation-based control of reconfigurable robotic workcells (Reconcell)
ElMoaqet [29]	Testbed	IoT, CPS	Development of a testbed for I4.0
Saez et al. [30]	Testbed	SM	Hybrid simulation for real-time performance monitoring
Tang et al. [31]	Proof-of-concept	SF, SM, BD	A cloud-assisted self-organized architecture for distributed manufacturing scheduling
Wang et al. [32]	Proof-of-concept	IoT, SM, CC	Cloud-based smart manufacturing for personalized consumption demands
Chen et al. [33]	Testbed	IoT, BD, CC	Hierarchical architecture of smart factory
Schluse et al. [34]	Testbed	SM	Experientable Digital Twin of an autonomous robotic workcell (ReconCell)
ElMaraghy et al. [35]	Testbed	SM	Reconfigurable manufacturing systems. Introduction of new product family
Nie et al. [36]	Testbed	SM	Reference model of an I4.0 demonstration production line using Ptolemy II
Masoudinejad et al. [37]	Testbed	IoT, CPS, AGV	Cellular transport systems in material handling and warehouse applications
Atorf et al. [38]	Testbed	SM	Framework for model configuration and simulation-optimization of a ReconCell
Liu et al. [39]	Testbed	CPS, CC	Development of the Cyber-Physical Manufacturing Cloud architecture
Li et al. [40]	Proof-of-concept	SF, SM, BD	Distributed manufacturing scheduling
Zarte et al. [41]	Testbed	SM	Predictive maintenance using AnyLogic
Lu et al. [42]	Testbed	CPS, AGV	RFID-enabled positioning approach in AGV for smart factories
Giorgio et al. [43]	Proof-of-concept	VR	Human-robot collaboration
Okuda et al. [44]	Testbed	IoT, CC	Development of an IoT testbed. Controlled experiment using OPC UA protocol
Wan et al. [45]	Testbed	IoT, CPS, BD	IIoT architecture to operate physical devices and provide an interface for data exchange
Wan et al. [46]	Proof-of-concept	IoT, CC	Mobile services for customization of intelligent manufacturing systems

• Search string: TOPIC: "Industry 4.0" AND TOPIC: testbed OR "proof-of-concept")
• Legend: SM—Simulation; SF—Smart Factory; VR—Virtual Reality; AR—Augmented Reality; BD—Big Data; CC—Cloud Computing

In Li et al. [40], the authors proposed a smart manufacturing framework of I4.0, modeling the shop-floor-entities as a multi-agent system (MAS). They also proposed an evaluation and control algorithm to reduce the load unbalance in manufacturing scheduling with big data feedback assistance. Their model consists mainly of four agent types, i.e. task agent (TA), smart product agents (PA), machining agents (MA), and conveyor agents (CA) that compete and cooperate in the self-organized manufacturing system. To validate their framework, the authors conducted a proof-of-concept experiment in a learning factory setting with a candy-picking process at the South China University of Technology, assuming that nine types of mixed candies need to be packed. The candy-picking process is mainly composed of five servers (for private cloud assistance), five robots (representing the MAs), and five conveyor s controlled through PLCs (representing the CAs). Their findings suggest that their control method can improve both load balance and efficiency. This candy-pinking process is also used as a validation setting in other similar studies [31–33, 45, 46].

I4.0 systems must be flexible and reconfigurable to allow highly customized products [1]. In [27], the authors describe in detail the development of a modular factory that serves as a testbed to analyze reconfigurable manufacturing systems. The main aspects investigated to speed factory configuration are self-layout recognition, fast workstation and robot programming, interlayer information sharing, and configurable software for shop-floor monitoring. Composed of ten workstations and multiple components (e.g. robotic arm, pallet transfer system, NFC tags, infrared communication modules), the testbed adopts a three-layer MAS architecture with three agent types (i.e., coordinator, workstation agent, workstation executor) for all shop floor decisions. Another modular learning factory used as a testbed for research, to analyze the introduction of a new product family in reconfigurable systems, is reported in [35]. According to the authors, the modular and changeable learning factory (iFactory), located at the University of Windsor, Canada, is equipped with robotic and manual assembly stations, a computer vision inspection station, an automated storage and retrieval system, an RFID system, and several material handling modules.

In Zarte et al. [41], the authors used an educational manufacturing system at the University of Applied Sciences Emden/Leer, in Germany, as a testbed to evaluate their predictive maintenance simulation model to improve production planning. Their simulation framework is based on a MAS and was implemented using the software AnyLogic. The physical structure of the laboratory, also described in Zarte et al. [47], includes an input/output module to feed raw material and deliver finished products, a camera, a high-rack warehouse, a main conveyor belt line, three industrial robots, two workplaces, and multiple buffer positions for the carriers to assembly three products, all made of LEGO Duplo bricks equipped RFID tags.

In Saez et al. [30], the authors present a hybrid simulation framework for shop-floor real-time performance monitoring by combining discrete event simulation (DES) with continuous dynamics (CD) in a sole simulation environment. The manufacturing testbed at the University of Michigan, USA, mainly composed of two Fanuc robots, four CNC milling machines, and a conveyor loop, was used to validate their framework. The testbed was modeled in MATLAB/Simulink, and real data (both discrete and continuous signals) were gathered from the PLC using an IIoT agent, i.e. a Rockwell Automation Industrial Internet of Things (IIoT) adapter. The novelty of their framework is allowing the comparison of real-time shop-floor data with real-time simulation data to analyze system and machine performance. Another example for real time monitoring with predictive functionalities can be found in [48]. In [36], the authors used the open source platform Ptolemy II to model and simulate a heterogeneous system. Their main goal was to model and simulate the I4.0 demonstration production line located at Shenyang Institute of Automation, in China (that serves as a testbed for smart and intelligent manufacturing) to provide a reference model for its upgrading. The production cell counts with different workstations, conveyor s, pallets with RFID tags, robots, and AGVs. Their model applies mainly a DES approach [49].

Digital Twin is another key simulation technology in I4.0, used to converge manufacturing physical and digital spaces. In [34], the authors discuss the concept of experimentable digital twin s (EDTs) and their interconnection. According to the authors, EDTs combines digital twin s with model-based systems engineering and simulation technology. They developed a full digital representation of a reconfigurable assembly work cell (named ReconCell), equipped with two UR10 robots (cobots). They also employed the concept of a virtual testbed to allow the network of interacting EDTs. The ReconCell also appears as an environmental setting in other scientific studies [28, 38]. In [24], the authors proposed a general framework for knowledge-driven digital twin manufacturing cell (KDTMC) for intelligent manufacturing (characterized by autonomy and self-optimization). The KDTMC is based on three main technologies (i.e. digital twin, dynamic knowledge bases, knowledge-based intelligent skills) operating in four spaces (i.e. physical, digital, knowledge, social), endowed with the capacity of self-thinking, self-decision-making, self-execution, and self-improving. To show KDTMC feasibility and provide practical insights, the authors built a testbed using Xi'an Jiaotong University's lab and presented three application examples concerning intelligent process planning, scheduling, analysis, and dynamic regulation. The physical system consists of two robotic arms, three machine tools, two buffers, one AGV, a warehouse, and a Kanban system.

There are other I4.0 research testbeds documented in the literature. In [39], the authors introduce the concept of cyber-physical manufacturing cloud (CPMC) to allow direct operation and monitoring of machines remotely in a manufacturing cloud. They also present a scalable service-oriented layered architecture of CPMC to allow manufacturing web services and cross-platform applications. Moreover, the authors describe the development of a testbed based on CPMC architecture, which links two geographically distributed manufacturing sites over the Internet. One site is located at the University of Arkansas, in the USA, equipped with one CNC, two 3D printers, two robotic arms, three controllers, and one local server. The other site is located at the Missouri Institute of Science and Technology, also in the USA, equipped with one 3D printer, one robotic arm, two controllers, and one local server. In [37], the authors describe a real-scale IoT-based warehouse testbed (PhyNetLab), used to evaluate logistic solutions in material handling and warehouse applications. The PhyNetLab includes 50 cellular transport system (mobile robots for materials handling), operating as CPS nodes, intelligent bins, and intelligent containers. In [42], the authors used simulations and a smart factory testbed to evaluate the practicality of their proposed RFID-enabled positioning approach in AGV system for smart factories. In [29], the authors describe the development of a cyber-physical system of I4.0 testbed (CPSI4.0), a small serial line equipped with several sensors, a conveyor belt, an RFID system, and an IoT Gateway to make data available at the local webserver. The testbed is mainly directed to research labs as well as for education and training purposes, with a concentration on sensor technology and measurement data acquisition. Similarly, in [44], the authors developed a small IoT testbed and performed experiments, including OPC UA and database on the cloud. Other papers adopt virtual environments to develop a testbed [25] or proof-of-concept [26, 43].

7.3 Lab Description

The learning factory of Produtique Québec , located at Sherbrooke, in the Canadian province of Quebec, was inaugurated in 2009 to support education, training, and applied research in mechatronics and industrial informatics. At that time, the lab also counted with one of the most advanced 3D printers in the province of Quebec, as additive manufacturing was emerging as a key technology. The I4.0 became a major topic in the manufacturing world after 2011, along with its primary enabling technologies, i.e. IoT, CPS, and Smart Factory [1]. Hence, the focus of the learning factory shifted towards I4.0 concepts. In 2018 Produtique Québec joined forces with the Cégep de Sherbrooke, Université de Sherbrooke, and other partners to create the CEGEMI , gradually transforming the learning factory into a living lab. They aim to support education, training, research, and innovation as well as to assist manufacturing SMEs in their digital transformation by offering various services, including managerial and technical assistance. Even though, similar to SMEs [5] and other I4.0 learning factories [14, 15], its transition to I4.0 will be incremental, by incorporating key I4.0 enabling technologies, such as IoT, Cyber-Physical Production System (CPPS), the Internet of Services (IoS), and simulation.

Figure 7.2 shows an overview of the lab's current structure. The process has multiple stations and counts mainly with a UR5 collaborative robotic arm (cobot), an LR Mate robotic arm, a CNC turning center, and a CNC milling center to produce didactically prepared products. Concerning education and training, the Cégep de Sherbrooke and the Université de Sherbrooke have done multiple projects to assist their academics, including the lab as a part of their curricula. Programming robots to perform specific tasks and the automation of the assembly line are some of the didactic topics used to develop specific teaching-learning situations. In the last academic term, around 50 students worked with projects to learn how to program robotic arms. Starting from the next academic term, Masters students from the Université de Sherbrooke will use the lab study the application of new management principles considering the cobot in the production line. It also served to demonstrate the use of some I4.0 technologies to SME (e.g. IIoT, cobots). Current research in the lab includes a project to calculate Overall Equipment Effectiveness (OEE) in real

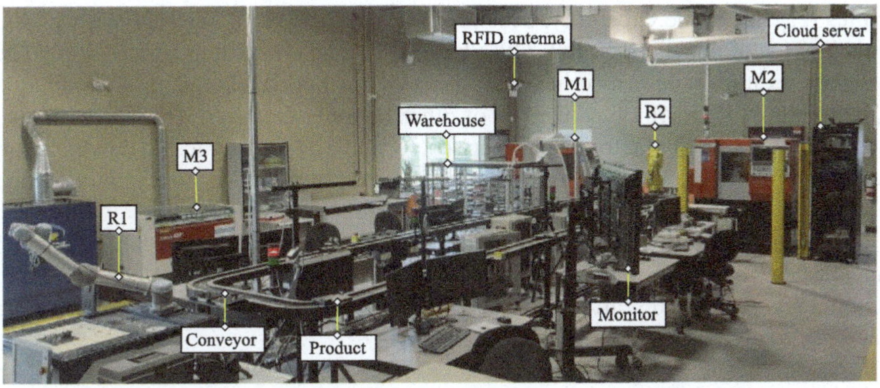

Legend: R1- Robot UR5 (Universal Robot); R2 - Fanuc LR Mate iC 5L Robotic Arm; M1 - EMCO Concept Mill 155 Vertical Machining Centre; M2 - EMCO Concept Turn 250 Turning Center; M3 - Laser Cutting Machine.

Fig. 7.2 Lab infrastructure overview (*Source* Author)

time using a Fanuc robotic arm through the advanced internet data gateway EWon Flexy205; a project to develop a social media platform for collaborative data sharing; a project about achieving the ISA 95 standard to enable the integration of enterprise and control systems; among others. Currently, there are 3 Ph.D., 1 M.Sc. student and several research professionals using the lab to develop their research.

7.4 Discussion and Conclusions

Numerous research papers adopting a learning factory to develop an I4.0 testbed or proof-of-concept were identified and analyzed in Sect. 7.2.1, demonstrating the relevance of learning factories in supporting I4.0 applied research. These articles also reveal research opportunities areas to Produtique Québec and CEGEMI lab, such as predictive maintenance, distributed production planning and control, and different modeling and simulation approaches in I4.0 (e.g. agent-based, discrete-event, hybrid simulation, digital twin s, virtual reality, augmented reality). They also provide different examples of I4.0 applications, methods, and technologies, which are extendable to the lab. Moreover, Produtique Québec and CEGEMI lab present a similar physical infrastructure to that of other learning factories documented in the scientific literature, such as in [24, 30, 41]. Therefore, its roadmap towards I4.0 can be more easily defined. Furthermore, it reinforces the potential of the lab for the development of applied research in different areas of I4.0.

This study reveals an increasing number of publications on I4.0 learning factories. It also describes how learning factories can support applied research in I4.0. To this end, studies with a testbed or a proof-of-concept experiment developed in learning factories were identified and analyzed. These studies cover different I4.0 design principles (e.g. real-time capability, decentralization, modularity, interoperability) and enabling technologies (e.g., IoT, CPS, big data, cloud computing, simulation). The practical examples described here may help the further development of Produtique Québec and CEGEMI lab as well as of other learning factories. This research also suggests that learning factories may support technology transfer to SME, which also tends to move incrementally towards digitalization by showing how to incorporate and testing a wide range of I4.0 methods and technologies.

Future research may consider Produtique Québec and CEGEMI lab to develop an I4.0 testbed or proof-of-concept experiment using a modeling and simulation technology. Another research avenue is the integration of the lab with other learning factories to create a distributed supply chain.

Acknowledgements This research is supported by the Federal Institute of Education, Science and Technology of Sao Paulo (IFSP) in Brazil, by Mitacs Globalink, by Quebec's Ministry of Economy and Innovation, and by the Natural Sciences and Engineering Research Council (NSERC) of Canada [grant numbers RGPIN-2018-06680].

References

1. Kagermann, H., Helbig, J., Hellinger, A., et al.: Recommendations for implementing the strategic initiative industrie 4.0: securing the future of German manufacturing industry; final report of the Industrie 4.0 Working Group. Forschungsunion, Frankfurt (2013)
2. Liao, Y., Deschamps, F., Loures, E.F.R. et al.: Past, present and future of Industry 4.0 - a systematic literature review and research agenda proposal. Int. J. Prod. Res. (2017). https://doi.org/10.1080/00207543.2017.1308576
3. Schlechtendahl, J., Keinert, M., Kretschmer, F. et al.: Making existing production systems Industry 4.0-ready. Prod. Eng. (2015). https://doi.org/10.1007/s11740-014-0586-3
4. Illa, P.K., Padhi, N.: Practical guide to smart factory transition using IoT. In: IEEE Access, Big Data and Edge Analytics (2018). https://doi.org/10.1109/ACCESS.2018.2872799
5. Moeuf, A., Pellerin, R., Lamouri, S. et al.: The industrial management of SMEs in the era of Industry 4.0. Int. J. Prod. Res. (2018). https://doi.org/10.1080/00207543.2017.1372647
6. Abele, E., Chryssolouris, G., Sihn, W., et al.: Learning factories for future oriented research and education in manufacturing. CIRP Ann. Manuf. Techn. (2017). https://doi.org/10.1016/j.cirp.2017.05.005
7. Abele, E., Metternich, J., Tisch, M., et al.: Learning factories for research, education, and training. CIRP Ann. Manuf. Techn. (2015). https://doi.org/10.1016/j.procir.2015.02.187
8. Tisch, M., Hertle, C., Abele, E., et al.: Learning factory design: a competency oriented approach integrating three design levels. Int. J. Compt. Integ. M (2016). https://doi.org/10.1080/0951192X.2015.1033017
9. Okoli, C., Schabram, K.: A guide to conducting a systematic literature review of information systems research. Wkng Papers Inf. Sys. (2010). https://doi.org/10.2139/ssrn.1954824
10. Ghobakhloo, M.: The future of manufacturing industry: a strategic roadmap toward Industry 4.0. J. Manuf. Technol. Manag. (2018).https://doi.org/10.1108/JMTM-02-2018-0057
11. De Souza Dutra, M.D., Da Conceição, Júnior G., De Paula, Ferreira W., et al.: A customized transition towards smart homes: a fast framework for economic analyses. Appl. Energ. (2020). https://doi.org/10.1016/j.apenergy.2020.114549
12. Wienbruch, T., Leineweber, S., Kreimeier, D. et al.: Evolution of SMEs towards Industrie 4.0 through a scenario based learning factory training. Procedia Manuf. (2018). https://doi.org/10.1016/j.promfg.2018.04.007

13. Elbestawi, M., Centea, D., Singh, I., et al.: SEPT learning factory for industry 4.0 education and applied research. Procedia Manuf. (2018). https://doi.org/10.1016/j.promfg.2018.04.025
14. Karre, H., Hammer, M., Kleindienst, M. et al.: Transition towards an Industry 4.0 state of the LeanLab at Graz University of Technology. Procedia Manuf. (2017). https://doi.org/10.1016/j.promfg.2017.04.006
15. Baena, F., Guarin, A., Mora, J. et al.: Learning factory: the path to industry 4.0. Procedia Manuf. (2017). https://doi.org/10.1016/j.promfg.2017.04.022
16. Nardello, M., Madsen, O., Moller, C.: The smart production laboratory: a learning factory for industry 4.0 concepts. In: Proceedings of Perspectives in Business Informatics Research, Aalborg University, Copenhagen, 28–30 Aug 2017
17. Wank, A., Adolph, S., Anokhin, O. et al.: Using a learning factory approach to transfer industrie 4.0 approaches to small-and medium-sized enterprises. Procedia CIRP (2016). https://doi.org/10.1016/j.procir.2016.05.068
18. ElMaraghy, H., ElMaraghy, W.: Learning integrated product and manufacturing systems. Procedia CIRP (2015). https://doi.org/10.1016/j.procir.2015.02.222
19. Erol, S., Jäger, A., Hold, P. et al.: Tangible Industry 4.0: a scenario-based approach to learning for the future of production. Procedia CIRP (2016). https://doi.org/10.1016/j.procir.2016.03.162
20. Tisch, M., Metternich, J.: Potentials and limits of learning factories in research, innovation transfer, education, and training. Procedia Manuf. (2017). https://doi.org/10.1016/j.promfg.2017.04.027
21. Fortier, P., Michel, H.: Computer Systems Performance Evaluation and Prediction. Digital Press, Burlington (2003)
22. Kendig, C.E.: What is proof of concept research and how does it generate epistemic and ethical categories for future scientific practice? Sci. Eng. Ethics (2016). https://doi.org/10.1007/s11948-015-9654-0
23. Neto, A.J.R., Borges, M.M., Roque, L.: A preliminary study of proof of concept practices and their connection with information systems and information science. In: Proceedings of the Sixth International Conference on Technological Ecosystems for Enhancing Multiculturality, University of Salamanca, Salamanca, 24–26 Oct 2018
24. Zhou, G., Zhang, C., Li, Z., et al.: Knowledge-driven digital twin manufacturing cell towards intelligent manufacturing. Int. J. Prod. Res. (2019). https://doi.org/10.1080/00207543.2019.1607978
25. Delbrugger, T., Meißner, M., Wirtz, A., et al.: Multi-level simulation concept for multidisciplinary analysis and optimization of production systems. Int. J. Adv. Manuf. Tech. (2019). https://doi.org/10.1007/s00170-019-03722-1
26. Tzimas, E., Vosniakos, G.C., Matsas, E.: Machine tool setup instructions in the smart factory using augmented reality: a system construction perspective. Int. J. Interact. Des. Manuf. (2019). https://doi.org/10.1007/s12008-018-0470-z
27. Kim, D.Y., Park, J.W., Baek, S., et al.: A modular factory testbed for the rapid reconfiguration of manufacturing systems. J. Intell. Manuf. (2019). https://doi.org/10.1007/s10845-019-01471-2
28. Priggemeyer, M., Rossmann, J.: Simulation-based control of reconfigurable robotic workcells: interactive planning and execution of processes in cyber-physical systems. In: 50th International Symposium on Robotics, Messe München, Munich, 20–21 June 2018
29. ElMoaqet, H., Ismael, I., Patzolt, F. et al.: Design and integration of an IoT device for training purposes of industry 4.0. In: Proceedings of the 2nd International Symposium on Computer Science and Intelligent Control, Stockholm University, Stockholm, 21–23 Sep 2018
30. Saez, M., Maturana, F.P., Barton, K., et al.: Real-time manufacturing machine and system performance monitoring using internet of things. IEEE Trans. Autom. Sci. Eng. (2018). https://doi.org/10.1109/TASE.2017.2784826
31. Tang, H., Li, D., Wang, S. et al.: CASOA: an architecture for agent-based manufacturing system in the context of industry 4.0. IEEE Access (2018). https://doi.org/10.1109/ACCESS.2017.2758160

32. Wang, S., Wan, J., Imran, M., et al.: Cloud-based smart manufacturing for personalized candy packing application. J. Supercomput. (2018). https://doi.org/10.1007/s11227-016-1879-4
33. Chen, B., Wan, J., Shu, L. et al.: Smart factory of industry 4.0: key technologies, application case, and challenges. IEEE Access (2017). https://doi.org/10.1109/ACCESS.2017.2783682
34. Schluse, M., Priggemeyer, M., Atorf, L. et al.: Experimentable digital twins streamlining simulation based systems engineering for industry 4.0. IEEE Trans. Ind. Inform. (2018). https://doi.org/10.1109/TII.2018.2804917
35. ElMaraghy, H., Moussa, M., ElMaraghy, W., et al.: Integrated product/system design and planning for new product family in a changeable learning factory. Procedia Manuf. (2017). https://doi.org/10.1016/j.promfg.2017.04.008
36. Nie, Z., Wang., Zeng, P. et al.: Modeling industry 4.0 demonstration production line using Ptolemy II. In: 43rd Annual Conference of the IEEE Industrial Electronics Society, Beijing, 29 Oct–1 Nov 2017
37. Masoudinejad, M., Venkatapathy, A.K.R., Emmerich, J. et al.: Smart sensing devices for logistics application. In: Proceedings of Sensor Systems and Software, Sophia Antipolis, Nice, 1–2 Dec 2016
38. Atorf, L., Schorn, C., Rossmann, J. et al.: A framework for simulation-based optimization demonstrated on reconfigurable robot workcells. In: Proceedings of the IEEE International Systems Engineering Symposium (ISSE), Vienna, 11–13 Oct 2017
39. Liu, X.F., Shahriar, M.R., AlSunny, S.N., et al.: Cyber-physical manufacturing cloud: architecture, virtualization, communication, and testbed. J. Manuf. Sys. (2017). https://doi.org/10.1016/j.jmsy.2017.04.004
40. Li, D., Tang, H., Wang, S. et al.: A big data enabled load-balancing control for smart manufacturing of industry 4.0. Cluster Comput. (2017). https://doi.org/10.1007/s10586-017-0852-1
41. Zarte, M., Wunder, U., Pechmann, A.: Concept and first case study for a generic predictive maintenance simulation in AnyLogic™. In: 43rd Annual Conference of the IEEE Industrial Electronics Society, Beijing, 29 Oct–1 Nov 2017
42. Lu, S., Xu, C., Zhong, R.Y., et al.: A RFID-enabled positioning system in automated guided vehicle for smart factories. J. Manuf. Syst. (2017). https://doi.org/10.1016/j.jmsy.2017.03.009
43. De Giorgio, A., Romero, M., Onori, M., et al.: Human-machine collaboration in virtual reality for adaptive production engineering. Procedia Manuf. (2017). https://doi.org/10.1016/j.promfg.2017.07.255
44. Okuda, M., Mizuya, T., Nagao, T.: Development of IoT testbed using OPC UA and database on cloud. In: 56th Annual Conference of the Society of Instrument and Control Engineers of Japan (SICE), Kanazawa University, Kanazawa, 19–22 Sep 2017
45. Wan, J., Tang, S., Shu, Z. et al.: Software-defined industrial internet of things in the context of industry 4.0. IEEE Sens. J. (2016). https://doi.org/10.1109/JSEN.2016.2565621
46. Wan, J., Yi, M., Li, D. et al.: Mobile services for customization manufacturing systems: an example of industry 4.0. IEEE Access (2016).https://doi.org/10.1109/ACCESS.2016.2631152
47. Zarte, M., Pechmann, A., Wermann, J., Gosewehr, F., Colombo, A.W.: Building an industry 4.0-compliant lab environment to demonstrate connectivity between shop floor and IT levels of an enterprise. In: 42nd Annual Conference of the IEEE Industrial Electronics Society, Palazzo dei Congressi, Florence, 23–26 Oct 2016
48. Eiskop, T., Snatkin, A., Karjust, K.: Production monitoring system with predictive functionality. Int. J. Eng. Sci. Technol. **12**(9), 2410–2425 (2017)
49. Da Costa, L.S., Lucio, W.D.S., Da Silva, A.M., et al.: Discrete simulation applied to the production process of electronic components. Indep. J. Manag. Prod. (2017). https://doi.org/10.14807/ijmp.v8i5.603
50. De Oliveira, C.A.S., De Paula, Ferreira W., Mendes, R.C., et al.: Metaheuristic ILS with path relinking for the number partitioning problem. Indep J Manag Prod (2017). https://doi.org/10.14807/ijmp.v8i5.597

Part III
Machine Learning in Industry 4.0

As a technologist, I see how Artificial Intelligence and the fourth Industrial Revolution will impact every aspect of People's lives.

Fei-Fei Li
Stanford University

Chapter 8
Intelligent Digital Twin System in the Semiconductors Manufacturing Industry

Foaad Haddod and Alexiei Dingli

Abstract The introduction of Industry 4.0 and Internet of Things (IoT) in the manufacturing industry has widely contributed to improving the underlying processes. It will usher the doors towards the implementation of newer emerging technologies such as Smart Manufacturing and Digital Twin technologies (DT). This chapter highlights aspects of a new design for an intelligent system which involves the creation of a DT system within the semiconductor manufacturing area. The system will also include designing intelligent agents, we name them as Intelligent Digital Humming tool (IDH) tool. A design for a robust automated system is also proposed to provide a high level of convergence and synchronization between the developed DT and the physical manufacturing plant. On this basis, the chapter provides a new approach that is going to help an international semiconductor manufacturing company to digitize and automate their physical production process. Furthermore, this project aims to design a high-level application that allows engineers to facilitate their product design, improve their planning and incorporate predictive maintenance. This paper aims to overcome the heterogeneity, connectivity and automation challenges of designing an automated DT that controls the physical manufacturing processes. Also, a novel new Intelligent Digital Humming tool (IDH) is proposed to tackle the traditional digital twin automation challenges.

8.1 Introduction

The digitalization of the manufacturing industry fuelled the creation of sophisticated virtual models, which are referred to as Digital Twins (DTs) and this applies throughout all the stages of product realization [19]. Since its inception, DTs have been used in many fields and have demonstrated great potential. With the characteristics of

F. Haddod · A. Dingli (✉)
University of Malta, Msida, Malta
e-mail: alexiei.dingli@um.edu.mt

F. Haddod
e-mail: foaad.haddod.15@um.edu.mt

A. Dingli et al. (eds.), *Artificial Intelligence in Industry 4.0*,
Studies in Computational Intelligence 928,
https://doi.org/10.1007/978-3-030-61045-6_8

ultra-high synchronization and fidelity, convergence between the physical and virtual products is exhibiting high potential in product design, product manufacturing, and product service [21]. An Intelligent DT (IDT) for a production process enables the coupling of the production system with its digital equivalent which ensures the minimum delay between the time of data acquisition and the creation of the DT [23].

In the coming years, it is expected that an enriched big data environment incorporating smart manufacturing concepts such as IDT will further enable advanced analytics. However, it is expected that the need to incorporate subject matter expertise in solution design will remain [15]. Furthermore, the integration of data collection, planning algorithms and performance prediction of guaranteed cycle times with a high probability is still an open task [7]. Zhong et al. [29] stated that to implement intelligent manufacturing fully; it is essential to focus on; platform technologies such as networks, IoT, virtualization, service technologies and smart objects. Finally, Posada et al. [16] also stated that the scientific community, especially those in visual computing, will face new exciting fields of research linked to the challenges brought forth by the upcoming industrial revolution.

Nowadays, more and smarter technologies are being developed and tested by different manufacturers which will revolutionise the manufacturing industry as a whole. What is known as industry 4.0 has provided a platform to develop smart technologies that attracted the interest of academics and industry professionals alike [9]. Digitising and transforming manufacturing is becoming of interest to most manufacturing companies [17]. Driven by emerging technologies such as the Internet of Things (IoT), industrial big-data, DTs and Industry 4.0, the concept of smart manufacturing will continue to grow, and we expect that new technologies and methods will emerge in this domain. Such a system can collect, record, accumulate and comprehensively process all the data from product design until retirement [17]. They have been mostly used for fault diagnosis, cause-and-effect identification, predictive maintenance and improving performance analysis. In the area of product design and development [21] they have been used to determine how and what ways can be implemented in order to created effective real-time interaction between the virtual system and its physical twin. Industry 4.0 paradigms in manufacturing have also faced a number of challenges such as; ingesting data from multiple sources, handling different sources and different data formats. Another challenge is the convergence of the physical system and the gathered data [5]. The industry set generally includes a large number of operations and process. These are further enhanced through the installation of data sensors located in the various machines which measure and log a multitude of readings.

Predicting the future is not something easy, but it is a very desirable feature, especially when one considers that issues such as machine downtime are one of the largest sources of production loss [13]. With DT technology, this scenario changes because DT performs these calculations without the need of being prompted by anyone. It will continuously analyse the supply chain and quickly simulate the ideal and most efficient scenario for any particular job [28]. It does so by looking at the current production, predicts potential downtime, slot in new tasks in the pipeline and makes suggestions based upon deadlines such as delivery targets and material available all at the click of a button. It is the ultimate aim of the DT to ensure that the production plant is optimised at all time, and it is continually readjusting those optimisation parameters in real-time based upon changes which occur on the factory floor. Thus, allowing factory plants to organise predictive maintenance, shifting the management's strict dependency on planned events, and allowing it to take real-time action from actual events [6]. Furthermore, the proposed system will not only predict a potential failure, but it will also help management identify the root cause of the issue. Besides, some of the failures will be eliminated from a production cycle once and for all. While others will be handled on time. In so doing, unplanned downtime can be massively reduced or even eliminate in some cases, thus maximising profitability and equipment reliability.

This chapter consists of the following parts; The next section is the literature review that includes background information about the main themes directly related to the project, namely Smart manufacturing, digital twin, and Industry 4.0. The following section highlights the system design, which provides details about the platform used and the big data methodologies adopted. Finally, it will delve in the results and discuss the various observations before concluding and explaining the next stage of the project.

8.2 Related Research

8.2.1 Industry 4.0

Industry 4.0 is a concept that refers to the fourth industrial revolution enabled by the use of the Internet of Things (IoT) technologies to create smart products, a smart production, and smart services. The use of Industry 4.0 applications holds an excellent opportunity to realize sustainable industrial value creation on three dimensions: economic, social and environmental [20]. Industry 4.0 has surfaced as one of the most discussed concepts. Also, it has gained significant popularity in both academia and the industrial sector [4]. In manufacturing, Industry 4.0 significantly influences the production environment with radical changes in the execution of operations [18]. In order to successfully implement Industry 4.0 technologies and technologies, it is essential to secure and effectively handle data and protect cross-company communication against attacks [24]. In manufacturing and production, the purpose of the Industry 4.0 initiatives is to bring together IT technologies and production processes to enable the emergence of smart, connected manufacturing [12]. In general, Industry 4.0 consists of several technologies and methods connected to the Internet leading to improved overall quality, productivity and its integration can be categorized as; horizontal integration, vertical integrations and end-to-end engineering integration [14].

8.2.2 Smart Manufacturing

Smart manufacturing (SM) refers to the use of smart technologies within manufacturing industry. It is considered as an improved version of intelligent manufacturing (IM), reflecting the magnitude and impact of smart technologies such as the Internet of Things, Cloud Computing, Cyber-Physical Systems and Big Data [26]. It refers to the use of advanced data analytics to improve system performance and decision-making [25]. Smart manufacturing system includes two main elements; data management and analysis. These provide real-time actionable information that can be used to optimize and build system intelligence into manufacturing operations [10]. With the advances in new-generation information technologies, especially big data and DT, smart manufacturing is becoming the focus of global manufacturing transformation and upgrading [17].

8.2.3 The Concept of DTs

The term DT is not new, and it was first introduced by Dr Michael Grieves, in 2003, in his Executive Course on Product Management Lifecycle (PML) at the University of Michigan in the United States [8]. What is new; however, is the recent interest in the digital twin concept and the applications that are fuelled by the emergence of a digitized manufacturing, cyber-physical systems, the need to gather large amounts of data and process such a vast collection of data sets [19]. Moreover, the concept of Industry 4.0 has also further impacted the visibility of DT applications within different sectors. A DT consists of three main parts; the physical product to be 'virtually twinned', the virtual product mimicking the physical product/twin and the connected data that ties the physical and virtual products (twins) together [21]. Thus, the main objective of a DT system is to design, test, manufacture and use the virtual version of the physical product, process, operation or system [8]. Since its introduction, the DT concept has been presented in several ways. For example, Boschert and Rosen [2], [p. 59] state that "the vision of the DT itself refers to a comprehensive physical and functional description of a component, product or system, which includes more or less all the information which could be useful in all—the current and subsequent—lifecycle phases". It is also defined as the creation of virtual models for physical objects in the digital world to simulate their behaviours [17]. The DT is a hierarchical system of mathematical models, computational methods and software services, which provides near real-time synchronization between the state of the real-world process or system and its virtual copy [1]. Furthermore, it is able to consistently provide all subsystems with the latest state of all the required information, methods and algorithms [3]. In a factory setting, a DT is about creating and maintaining a digital representation of the real world of the factory and supporting its management and reconfiguration by means of optimization and simulation tools, which are fed with real and updated factory data [11].

Based on these definitions, [21] identifies three main characteristics of a DT; real-time reflection; interaction and convergence; and self-evolution. Where Real-time reflection as a key feature of any DT system, any included communication between DT and physical twin systems should offer real-time interworking environments that provide an accurate reflection of a virtual model concerning physical asset and time synchronization between each simulator [27]. DT systems allow real-time monitoring of the status and progress of the physical 'twins' through real-time data collection, data integrations, and analysis [21]. On the other hand, the digital twin characteristic of interaction and convergence includes three main aspects; interaction and convergence in the physical space, interaction and convergence between historical data and real-time data; and interaction and convergence physical space and virtual space (the twins) [21]. Finally, the 'self-evolution' characteristic of digital twin refers to the ability to continuously update, in a real-time, the acquired data so that the virtual models of the digital twin system can be continuously improved by comparing the physical and virtual twin design and configurations [21, 22].

8.3 Proposed Work

This project proposes a new methodology for predictive maintenance within the microelectronics manufacturing industry driven by a state-of-the-art DT system. It aims to develop a DT system as a base of Imitation and Reinforcement learning that provides an autonomous predictive maintenance problem-solving in the physical semiconductor manufacturing facility. The developed DT system aims to merge various technologies and determine the best one among them that can be designed, developed, tested and implemented. Our proposed DT system is based on the operations of a manufacturing facility in one of the plants of an international semiconductor manufacturing company. The following subsections will provide details about the used approach and the developed system.

8.3.1 Microelectronics Manufacturing Plant Description

This project is based on a microelectronics manufacturing setting where there was a need to combine several manufacturing stages that are involved in the manufacturing process. Several manufacturing stages were grouped in a logical method based on time sequence and typical sequence of the production process. For this project, the entire production line was divided into three main stages, namely; Back-end, Assembly and Final testing process. Multiple processes were grouped under each of these three categorized production stages. Table 8.1 presents the number of sub-stages grouped under the three main stages and the number of the parameters under each one.

Table 8.1 Used parameters from the streamed data out of the semiconductor manufacturing processes

	Number of messages	Number of available parameters
EquipmentFailurs	1	21
EQuipmentStatus	50	276
Maintenance	2	18

For every considered main stage, several data messages were selected based on some pre-defined priorities. Each selected data message included many parameters that were collected from the production line in real-time. Some of the parameters used were considered as the most important ones, while others were discussed and selected by the company to provide an accurate indication of what could be the cause of some of the failures in the equipment. Also, all the data messages analysed included both; request events to ingest data from the machines and reply messages which included real-time data from the equipment status. Besides, the tested equipment had a small amount of real-time data due to the limited number of installed sensors. Thus, an IoT platform was installed to provide more data and measure more mechanical movements directly from the machines in the base stream. Such information will be visualised on the DT edge at a later stage.

8.3.2 Two-Way Connection Between the Physical Twin and the Digital Twin

In order to design effective connectivity between the physical and the digital twin, several different platforms were used. Since the data used was extensive, it was essential to design a high level of interconnectivity that provides smooth and real-time data flow between the physical and the digital twins which provide smooth and real-time interaction with the user. The uniqueness of the proposed system is its ability to process all the ingested data at the DT end. Two different data distribution systems were used to tackle the streamed data challenges and provide real-time data that is visualised into the DT edge. The KAFKA platform (https://kafka.apache.org) was used as a distributed storage system to structure all the acquired streamed data from the different production stages. Besides, KAFKA was also used to build a stream base process with fault tolerance features to provide high-quality streaming data to the DT edge. Figure 8.1 shows a diagram of the designed data acquisition system. It includes, from left to right, an IoT sensor tile system, an Equipment Integration Controller, a Tibco bus system and a KAFKA system to transfer data from the physical twin to the digital twin.

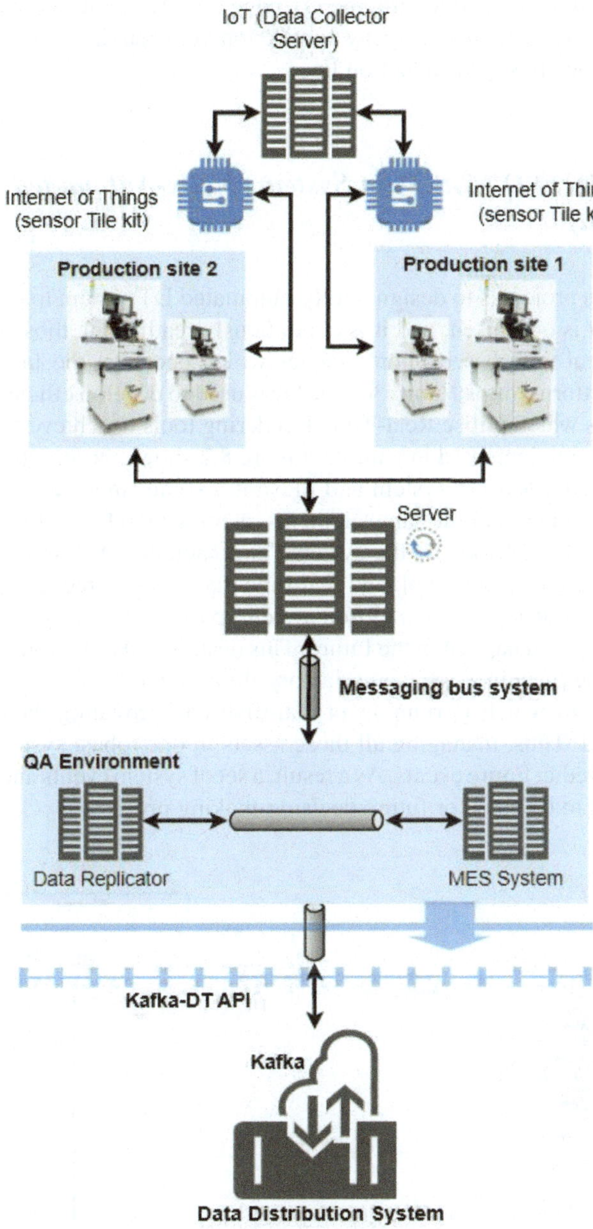

Fig. 8.1 Data acquisition infrastructure (*Source* self)

As can be seen from the diagram, the DT edge is fully connected with the physical system and have a robust two-way connection between them to achieve a 24/7 accessibility to the physical production line.

8.3.3 *Intelligent Digital Twin System (Fully-Automated Design)*

The scope of this project is to design a fully-automated DT system to ensure that the production plant is optimised, and it is continuously readjusting those optimisation parameters in real-time based upon changes which occur on the factory floor. A visualisation platform (https://unity3d.com) was used to develop efficient and realistic visualisations which utilise Real-Time Rendering tools to achieve several effects that are typically implemented in gaming. Figure 8.2 shows the structure of the first version of the developed DT system and illustrates some embedded elements. The designed system consists of the same elements mentioned in Fig. 8.1 but adding elements related to the DT side of the system. This includes; VR visualisation tool, Streaming Machine Learning tool, and real-time maintenance request system.

However, when dealing with a closed system, past events can give an excellent indication of what will happen in the future. This designed DT environment included 3D models of the machines and some factory floor elements that were considered as a closed system which is running in real-time and providing data that can be efficiently mined. Thus, managing all those assets in one robust system could help the DT system predict future events. As a result, a set of system events and instructions will be provided to be used for future decision-making processes.

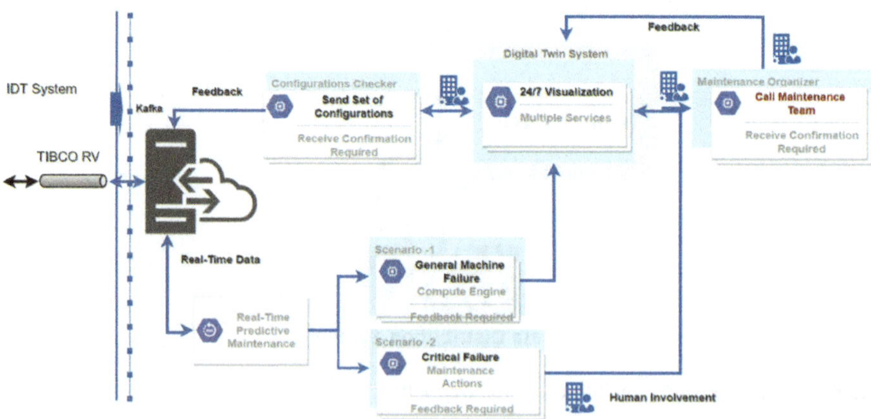

Fig. 8.2 Semi-automated digital twin design (*Source* self)

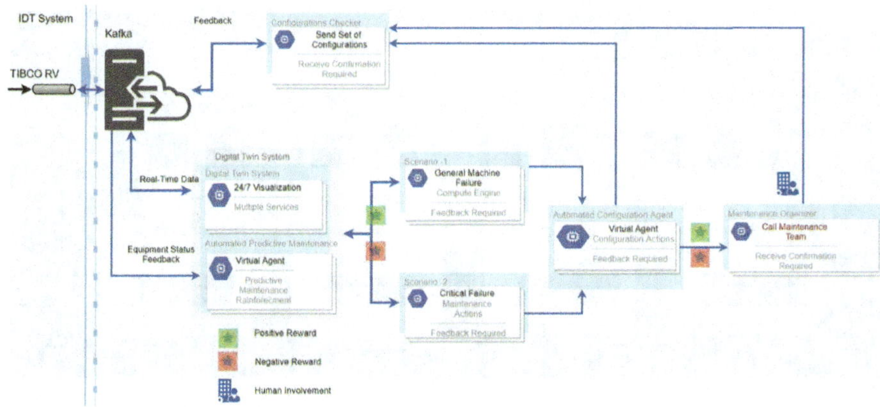

Fig. 8.3 Fully automated designed digital twin (*Source* self)

During the first phase of the project, it was clear that the developed DT system had an up-to-date overview of the monitored factory plant. Besides, a good number of factory floor actions were simulated in nearly real-time using the DT edge. We also noticed that several actions require human intervention which has caused unnecessary downtimes. A more automated improvement was provided to decrease equipment downtime. Reinforcement learning was used to improve the designed DT system's prediction capabilities. Figure 8.3 shows the second designed DT structure that includes; a virtual agent system, a maintenance system using reinforcement learning base, and maintenance request agent.

The proposed approach aims to reduce the amount of the collected data to make predictions about any potential equipment failures. The reason behind the second approach aimed to achieve some edge processing improvement. Besides, the prototype faced the challenge of designing a smooth intercommunication system that can tackle a large amount of ingested data. Also, the developed DT system required powerful hardware capabilities. Thus, using reinforcement learning will reduce the hardware requirement at the DT edge. Figure 8.4 shows a screen-shot from the developed DT system.

The next section provides an overview of the ongoing research that aims to develop a novel tool to improve the DT technology and enhance the DT system efficiency.

8.3.4 Intelligent Digital Humming Tools

In reality, processing all ingested streamed data out of a semiconductor manufacturing process is rather painful, mainly because of its volume. Therefore, designing a viable solution is a big challenge. At the time of writing this chapter, an initial version of the smart and intelligent tool has been designed. This tool aims to learn from a

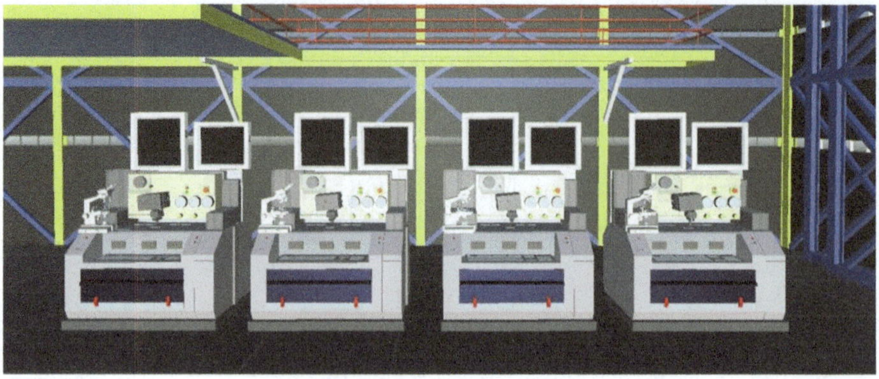

Fig. 8.4 The designed DT prototype showing various virtual machines which can also be found on the factory floor (*Source* self)

graphical environment so that it can control the DT system autonomously based on the understanding of what happened and the best reactions. A DT system interface was designed using Virtual Reality to create a realistic graphical environment, technology.

This section provides a general overview of the tool that acts as a triplet for the traditional DT systems and pushes its boundaries. The idea here is to design a smart, light, and intelligent tool that can learn from the designed graphical environment and to predict possible equipment failures. At the same time, this tool will automate different actions without the need of being prompted by anyone. The proposed tool can be flexible and sufficient to enhance different processes, for instance; quickly simulate the ideal and most efficient scenario for any particular job. It does so by looking at the real-time production behaviour, predicts potential downtime, slot in new tasks in the pipeline and makes suggestions based on deadlines such as delivery targets and the material available. Figure 8.5 shows a high-level diagram of the proposed Intelligent Digital Humming (IDH) scenarios.

The proposed IDH tool uses learning by 'imitation' technique to train itself on different failure events. All real-time failures are visualised into the designed DT system in an accurate and advanced visualisation tool. Training will allow the developed IDH to figure out the best machines' configurations that result in the best productivity. Once an IDH agent explores an optimal machine configuration, it will automatically reset the other equipment (from the same production line) which is producing the same products. Thus, there is no need for human intervention to figure out such insights and this results in reducing downtime while applying any reset operations. Figure 8.6 illustrates a scenario of learning and automating reconfiguration process using the proposed IDH tool. It also shows how the IDH tool learns from the behaviour of the designed DT assets using imitation technique. The more this tool learns, the higher the rewards it gets until it reaches 100. At this point, the IDH tool will be considered as fully competent and can control the operations of the

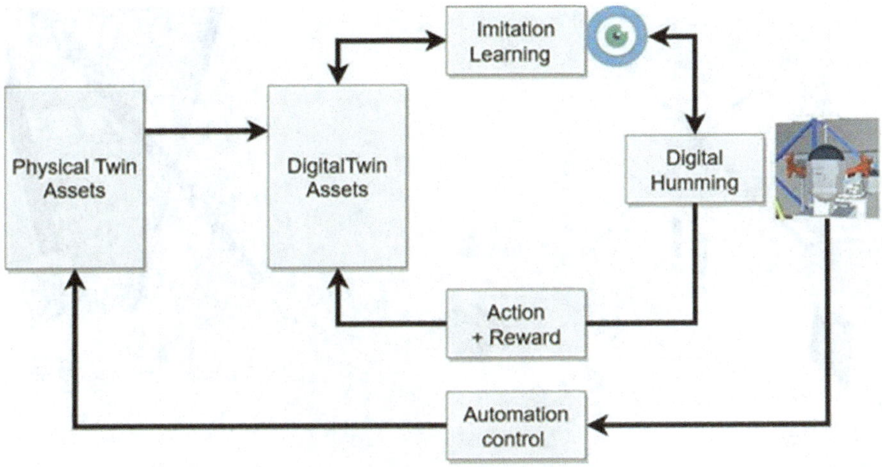

Fig. 8.5 Proposed intelligent digital humming tool learning scenario (*Source* self)

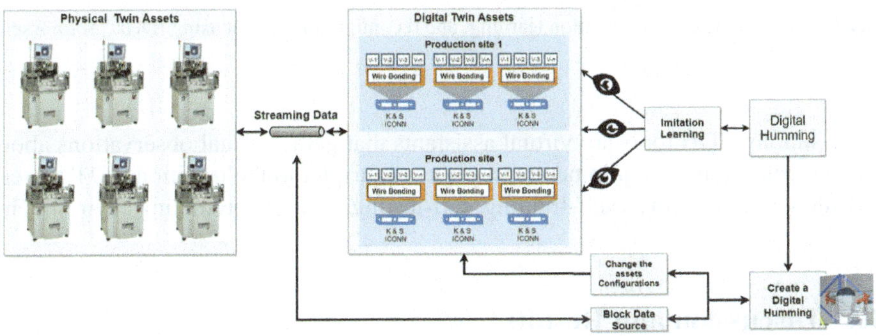

Fig. 8.6 Proposed intelligent digital humming (IDH) automated machine configuration scenario (*Source* self)

DT equipment. Furthermore, and after all the DT equipment being reconfigured, the IDH tool blocks the data source and starts to control the physical equipment.

Furthermore, a virtual training lab was designed using 3D models for the physical equipment to be used during the training of the IDH tools before they are embedded in the primary DT system. Thus, the physical system will never be interrupted, and any invalid action on the production lines is avoided. The designed 3D models are fully-connected with the actual physical equipment and behave similarly as their physical twins in real-time. Thus, the IDH tools will have the opportunity to detect failures and actions anomalies. Figure 8.7 illustrates a virtual learning process and reconfiguration process of assets within the DT system and using the developed IDH tool.

Fig. 8.7 Failure detection, imitation learning, and reconfiguration event using IDHs (*Source* self)

To sum up, IDH tools are virtual assistants that gather visual observations about specific conditions of equipment and serve as a triplet to the traditional DT system and allows it to become a self-training, self-testing, and self-predicting environment.

8.4 Discussion and Results

In the semiconductor manufacturing, it is crucial to design an intelligent digital twin system that manages big data and leverages the interconnectivity of machines to reach the goal of intelligent, resilient and self-adaptable machines. Based on this, the proposed research provides useful insights into the applicability of using the intelligent digital twin technology within the semiconductor manufacturing industry. While examining the designed prototype, the applicability of adopting the developed DT system within semiconductor manufacturing was tested. Furthermore, intense attention was given to the processes and documented potential obstacles the manufacturing process may encounter. It has been noted that there are considerable opportunities to overcome several obstacles that such a complex manufacturing process encounters every day. Equipment could remain out of service for an extended period; thus, providing more downtime and resulting in more losses. From the historical data of the machine's failure events, there were some general machine's errors that are categorized as unknown errors.

Such errors could end up with unnecessary downtime that can be easily avoided by automating restarting machines instead of depending on the manual intervention of operators. On the other hand, some failure events usually require maintenance order requests. To overcome such errors, a whole set of maintenance procedures has to be designed which handles uneven down-time. The initial results have shown a high level of convergence between the physical twin and the Digital twin with full accessibility to all the available data. Using gaming techniques, we have managed to mimic natural physical and mechanical actions. However, there was some latency in the ingestion of data. Such delays can be resolved by using the IDH tool. So far, one scenario has been tested using two different machines (one old and one new) to explore and compare the performance and productivity of the two machines and to make the older machine perform in a similar capacity as the new one.

8.5 Conclusion

The designed intelligent digital twin using gaming techniques has provided a high level of convergence and synchronization between the developed DT and the physical production line of a semiconductor manufacturing plant. Also, the proposed IDH tools have shown significant improvements in the performance of the designed DT system. Gaming design is a field full of exciting visualization technologies that can be easily implemented to simulate natural physical and mechanical behaviours. Such accurate information can then be visualized into the DT system and used by the proposed triplet system. Thus, this will allow our proposed IDH tools to learn by trial and error from different scenarios. The proposed IDH tools can easily explore what is happening in the DT systems in real-time and learn from it without interrupting the physical production facilities. Besides, less data is required from the physical twin, and new and more efficient scenarios can be suggested. To sum up, the proposed intelligent DT system together with the triplet tool, which is currently being developed, will be imperative to enhance the process of manufacturing companies.

8.6 Future Work

The developed prototype of the IDH has shown promising results on automating the DT system. The next phase of the project is to investigate the possibility of widening the functions of the IDH tools. The second phase of the project aims to improve the proposed tool that can overcome the DT systems challenges includes; connectivity, edge processing, and hardware requirement. The system will be designed to make these IDH learn from the DT system's environment and predict future failure events automatically from the environment. Reinforcement learning will be to make the proposed tools to learn and explore better after being embedded in the DT system.

References

1. Borodulin, K., Radchenko, G., Shestakov, A., Sokolinsky, L., Tchernykh, A., Prodan, R.: Towards digital twins cloud platform: microservices and computational workflows to rule a smart factory. In: Proceedings of the 10th International Conference on Utility and Cloud Computing, pp. 209–210 (2017)
2. Boschert, S., Rosen, R.: Digital twin-the simulation aspect. Mechatronic Futures, pp. 59–74. Springer, Berlin (2016)
3. Brenner, B., Hummel, V.: Digital twin as enabler for an innovative digital shopfloor management system in the ESB logistics learning factory at Reutlingen- University. Procedia Manuf. **9**, 198–205 (2017)
4. Buer, S.V., Strandhagen, J.O., Chan, F.T.: The link between industry 4.0 and lean manufacturing: mapping current research and establishing a research agenda. Int. J. Prod. Res. **56**(8), 2924–2940 (2018)
5. Cemernek, D., Gursch, H., Kern, R.: Big data as a promoter of industry 4.0: lessons of the semiconductor industry. In: 2017 IEEE 15th International Conference on Industrial Informatics (INDIN), pp. 239–244. IEEE (2017)
6. Fernandes, M., Canito, A., Bolon, V., Conceicao, L., Praca, I., Marreiros, G.: Predictive maintenance in the metallurgical industry: data analysis and feature selection. In: World Conference on Information Systems and Technologies, pp. 478–489. Springer (2018)
7. Furmans, K., Liberopoulos, G., Rimmele, M., Zimmermann, O.: Review of models for large scale manufacturing networks. In: Proceedings 11th Conference on Stochastic Modeling of Manufacturing and Service Operations (SMMSO 2017), pp. 251–260 (2017)
8. Grieves, M.: Digital twin: manufacturing excellence through virtual factory replication. Web pages (2014)
9. Hermann, M., Pentek, T., Otto, B.: Design principles for industrie 4.0 scenarios. In: 2016 49th Hawaii International Conference on System Sciences (HICSS), pp. 3928–3937. IEEE (2016)
10. Kuhn, D., Ahrens, J.: Smart manufacturing for continuous, high? Technology glass production. In: 77th Conference on Glass Problems, vol. 38, p. 45. Wiley Online Library (2017)
11. Kuts, V., Modoni, G.E., Terkaj, W., Tahemaa, T., Sacco, M., Otto, T.: Exploiting factory telemetry to support virtual reality simulation in robotics cell. In: International Conference on Augmented Reality, Virtual Reality and Computer Graphics, pp. 212–221. Springer (2017)
12. Lalanda, P., Morand, D., Chollet, S.: Autonomic mediation middleware for smart manufacturing. IEEE Internet Comput. **21**(1), 32–39 (2017)
13. Luo, B., Wang, H., Liu, H., Li, B., Peng, F.: Early fault detection of machine tools based on deep learning and dynamic identification. IEEE Trans. Ind. Electron. **66**(1), 509–518 (2018). Intelligent digital twin system in the semiconductors manufacturing industry 15
14. Manogaran, G., Thota, C., Lopez, D., Sundarasekar, R.: Big data security intelligence for healthcare industry 4.0. Cybersecurity for Industry 4.0, pp. 103–126. Springer, Berlin (2017)
15. Moyne, J., Iskandar, J.: Big data analytics for smart manufacturing: case studies in semiconductor manufacturing. Processes **5**(3), 39 (2017)
16. Posada, J., Toro, C., Barandiaran, I., Oyarzun, D., Stricker, D., de Amicis, R., Pinto, E.B., Eisert, P., Dollner, J., Vallarino, I.: Visual computing as a key enabling technology for industrie 4.0 and industrial internet. IEEE Comput. Graph. Appl. **35**(2), 26–40 (2015)
17. Qi, Q., Tao, F.: Digital twin and big data towards smart manufacturing and industry 4.0: 360 degree comparison. IEEE Access **6**, 3585–3593 (2018)
18. Sanders, A., Elangeswaran, C., Wulfsberg, J.P.: Industry 4.0 implies lean manufacturing: research activities in industry 4.0 function as enablers for lean manufacturing. J. Ind. Eng. Manag. (JIEM) **9**(3), 811–833 (2016)
19. Schleich, B., Anwer, N., Mathieu, L., Wartzack, S.: Shaping the digital twin for design and production engineering. CIRP Ann. **66**(1), 141–144 (2017)
20. Stock, T., Seliger, G.: Opportunities of sustainable manufacturing in industry 4.0. Procedia CIRP **40**, 536–541 (2016)

21. Tao, F., Cheng, J., Qi, Q., Zhang, M., Zhang, H., Sui, F.: Digital twin-driven product design, manufacturing and service with big data. Int. J. Adv. Manuf. Technol. **94**(9–12), 3563–3576 (2018)
22. Tuegel, E.J., Ingraffea, A.R., Eason, T.G., Spottswood, S.M.: Reengineering aircraft structural life prediction using a digital twin. Int. J. Aerosp. Eng. **2011** (2011)
23. Uhlemann, T.H.J., Lehmann, C., Steinhilper, R.: The digital twin: realizing the cyber-physical production system for industry 4.0. Procedia CIRP **61**, 335–340 (2017)
24. Wallis, K., Kemmer, F., Jastremskoj, E., Reich, C.: Adaption of a privilege management infrastructure (PMI) approach to industry 4.0. In: 2017 5th International Conference on Future Internet of Things and Cloud Workshops (FiCloudW), pp. 101–107. IEEE (2017)
25. Wang, J., Ma, Y., Zhang, L., Gao, R.X., Wu, D.: Deep learning for smart manufacturing: methods and applications. J. Manuf. Syst. **48**, 144–156 (2018)
26. Yao, X., Zhou, J., Lin, Y., Li, Y., Yu, H., Liu, Y.: Smart manufacturing based on cyber-physical systems and beyond. J. Intell. Manuf. **30**(8), 2805–2817 (2019)
27. Yun, S., Park, J.H., Kim, W.T.: Data-centric middleware based digital twin platform for dependable cyber-physical systems. In: 2017 Ninth International Conference on Ubiquitous and Future Networks (ICUFN), pp. 922–926. IEEE (2017)
28. Zheng, Z., Zheng, L., Yang, Y.: A discriminatively learned CNN embedding for person reidentification. ACM Trans. Multimed. Comput. Commun. Appl. (TOMM) **14**(1), 1–20 (2017)
29. Zhong, R.Y., Xu, X., Klotz, E., Newman, S.T.: Intelligent manufacturing in the context of industry 4.0: a review. Engineering **3**(5), 616–630 (2017)

[illegible]

Chapter 9
Automatic Defect Recognition in Nonwovens Using Images and Metadata Analysis—A Deep Learning Approach

Marco Calderisi, Gabriele Galatolo, Ilaria Ceppa, Tommaso Motta, and Francesco Vergentini

Abstract The production of the nonwoven fabric is a particular process that requires to avoid any kind of defect and contamination. High-resolution camera systems and different image analysis techniques are currently in use. Nevertheless, the common procedure requires a manned intervention. In this work we propose, using both metadata and images of the defects provided by the vision system, a joined application of machine learning, and deep learning techniques to completely automate the severe defect image identification in order to obtain high-quality products and reduce machine stops. By means of Principal Component Analysis, Hotelling's T-Squared, and Q residuals we were able to use the metadata to identify and remove anomalous samples; then, by applying the Partial Least Squared/Discriminant Analysis we performed feature selection on the metadata variables. Metadata pre-processed in such way and related images are used as an input of a combination of two different sub-networks: a Convolutional Neural Network to analyze the images and a Multi-Layer Feedforward Neural Network to analyze the metadata.

M. Calderisi · G. Galatolo (✉) · I. Ceppa
Kode Srl, Pisa, Italy
e-mail: g.galatolo@kode-solutions.net

M. Calderisi
e-mail: m.calderisi@kode-solutions.net

I. Ceppa
e-mail: i.ceppa@kode-solutions.net

T. Motta · F. Vergentini
A.Celli Nonwovens S.p.A, Porcari, Italy
e-mail: t.motta@acelli.it

F. Vergentini
e-mail: f.vergentini@acelli.it

A. Dingli et al. (eds.), *Artificial Intelligence in Industry 4.0*,
Studies in Computational Intelligence 928,
https://doi.org/10.1007/978-3-030-61045-6_9

9.1 Introduction

Quality assurance is a very important part of any production process. In the field of non-woven production, in particular, it is fundamental to be certain that the manufactured reels of fabric do not present any kind of visual imperfection, as this would result in the rejection of the reels by the clients. A system that automatically detects such defects from images of the material would greatly improve the production rate and quality of any process, and for this reason, the problem has been widely studied.

The classical approach involves the extraction of a series of features from the images to use them as input for some kind of classification algorithm [1]. There is a huge amount of possible features, and the disadvantage of this approach lies in the need to identify which are the best suited to represent the image for the specific task at hand [2, 3].

In most recent years Image Analysis techniques experienced an impressive boost, due to the development of the Deep Learning approach [4–6]. Those techniques opened a new set of possibilities also in the field of industrial defect detection: convolutional neural networks (CNN) architectures allow for automatic extraction of features from images, guided by the classification task itself [7]. Several applications of these methods already exist in the field of industrial defect detection. In [8], convolutional neural networks were used to identify imperfections on steel surfaces, using photometric stereo images, whereas in [9] they were successfully utilized to detect defects on steel objects from images taken in a real production line. In [10], a simple convolutional network, consisting of only two layers, convolution and a soft-max one, was exploited to detect cracks in concrete images. The technique has also been applied to the problem of defect detection using videos, as reported in [11]: in this work, CNNs were used to analyze hours of videos of rail surfaces, and automatically extract features and recognize six different possible "states" the surface could be in.

The problem we had to solve was the improvement and automation of the detection of severe defects in reels of non-woven fabric. Nowadays, machines that produce these materials are equipped with a sophisticated vision system, consisting of several cameras located on top of the machine. Embedded within the cameras, there is an automated defect detection algorithm. This software is capable of detecting very quickly if the piece of fabric is seeing presents any kind of imperfection. The detection of defects is done by computing a series of features over the fabric pictures. Nevertheless, the vision system is usually not enough by itself to satisfy the requirements of non-woven fabric production. While being fast and accurate in detecting every imperfection in the fabric, the system does not take into account the severity of the defects: if every flaw detected by the vision system was to cause a machine stop and the disposal of that part of the fabric, the amount of wasted time and material would be enormous. For these reasons, the automatic detection of defects is followed by a manual inspection: a quality assurance operator visually inspects most of the pictures of the defects taken by the vision system and flags those that he considers

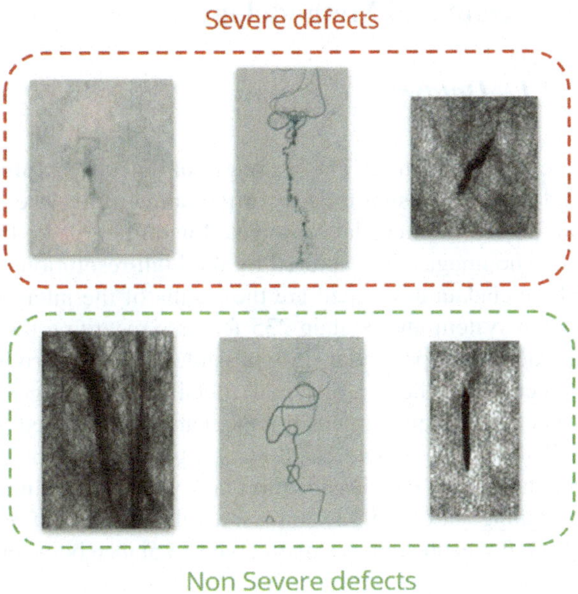

Fig. 9.1 Examples of severe/non-severe defects. Even if two images belong to the same group they may not have the same severity classification (*Source* Author)

being severe defects. Only these will be used to decide whether a reel should be discarded or not.

This manual process is clearly inefficient, as it involves human intervention and a visual inspection of as many pictures as possible. Moreover, the process is very error-prone, for various reasons. Firstly, the human operator can visualize only some pictures taken by the vision system, not all of them, so there will be some images that will be automatically classified as non-severe defects, just because the operator did not inspect them. Furthermore, the subjectivity of the quality assurance operator plays an important role in the determination of what is a severe defect and what isn't, which introduces misclassification errors also in the pictures he did inspect.

Our task consists in improving this phase of the defect detection system, building a classifier able to automatically distinguish between severe and non-severe defects. To do so, we use a dataset that consists of all the images taken by the vision system and the related metadata information, which is the set of features the system itself computes over the image. To determine if an image depicts a severe or a non-severe defect, we use the human operator evaluation of it, as this is the only information on the ground truth that is available to us. By doing so, we must take into account that we are introducing a certain degree of error in our model, as we might be labeling some of our data incorrectly. Moreover, the images in our dataset are very heterogeneous, as the vision system stores different versions of the same image under different lighting and positioning: in Fig. 9.1, examples of similar pictures that have different labeling. As shown, very similar images can have different labeling while there are cases where very different images are actually the same type of defect and share the same labeling.

9.2 Data and Methodology

9.2.1 Dataset

The dataset contains 71878 images of defects captured by the vision system at the end of the extrusion process of the nonwovens. The dataset contains the image of the defect; the metadata associated with the image; the classification of the operator. The images are captured by the high-resolution cameras of the vision system. The metadata associated are the results of the internal processing executed by the vision system and contain 235 features regarding the visual/physical properties of the defect. In particular, it is possible to find information on the dimension of the defect (width, height, diagonal), details of the image (contrast, the sharpness of the defect, background intensity), the features calculated by the system (in the metadata called $feat_1, \ldots, feat_n$ and $g_1, \ldots, g_n$), position on the roll, and so on. To prepare the dataset for the classification task we used a split of about 80/10/10 resulting in a training set of 59992 samples, a validation set of 5963 samples, and a test set of 5963. For each split, we maintained a balance between severe/non-severe defects.

9.2.2 Defect Classification Approach and Preprocessing

A naive approach to this task is to create a Convolutional Neural Network (CNN) to classify images in one of the two classes severe/non severe. The availability of the metadata, however, gives us the opportunity to also use this information within the classification process.

For this reason we prepared and test the solution with a single CNN, but also configurations which also take into account the metadata. To do so, we decided to use a combination of two networks: one of them is a CNN which processes the image; the other one is a multilayer feed-forward neural network that takes as an input the metadata associated to the image. The final layers of both networks are used as an input of a multilayer feed-forward neural network that provides the final classification of the sample.

The metadata were not used as an input as they are, but after a preprocessing stage that we split in two phases.

Firstly, we found and removed the outliers (**offline preprocessing**). In order to do that we applied the Principal Component Analysis on the metadata: on the resulting decomposition we used the confidence limits for Hotellings T-Squared and Q residuals [12] to remove from the training set images and metadata for those samples beyond the established limit of both statistics (Fig. 9.2).

Secondly, this procedure allows us to use the PCA decomposition to reduce the dimensionality of the metadata given as an input to the subnetwork which processes metadata (**online preprocessing**): the sample metadata, of dimension p as the number of features of the the metadata, are projected with the loadings matrix onto the new

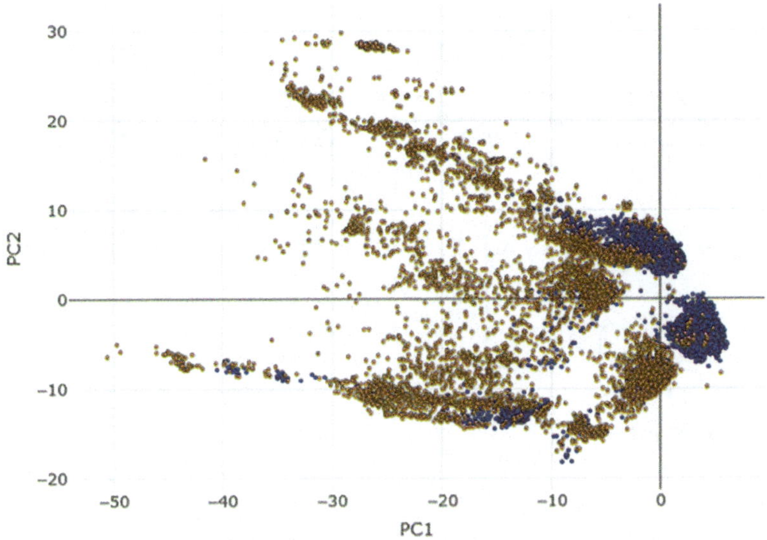

Fig. 9.2 Plot relative to the PCA executed on the metadata before the outliers removal. In yellow the samples identified as outliers (*Source* Author)

space, of dimension $k \ll p$, and the resulting projection is used as an input for the network.

As a further intermediate phase during the offline preprocessing, after the outliers removal, we can do a selection of the feature of metadata that mostly contribute to distinguish between severe and non severe defects. The feature selection has been performed using Partial Least Square/Discriminant Analysis. This classification method finds maximum covariances on the training set taking into account the labelling of each sample. To select the most important features, we used the variable input on projection (VIP) [13, 14], keeping only those variables with VIP value greater or equal to 1 [15, 16]. The metadata, for which only features with $VIP \geq 1$ are retained, are finally preprocessed as indicated previously (Fig. 9.3).

The final architecture we developed for the classification task is shown in Fig. 9.4. The subnetwork working on the defect's images uses a set of convolutional layers for features extraction, max pooling for downsampling, and dropout for regularization purposes, while the subnetwork which processes the metadata is basically a feed-forward neural network with two dense layers. The outputs of the last layers of both networks were concatenated to feed the final two layers feed-forward neural network, which gives the final classification.

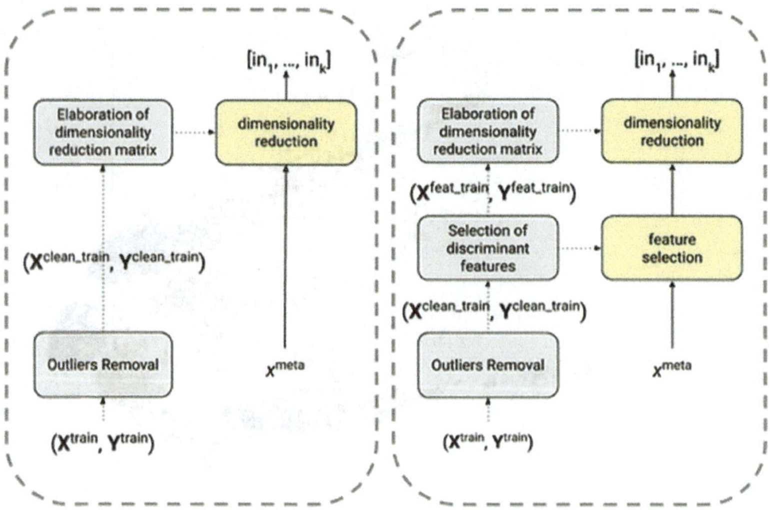

Fig. 9.3 In this image is shown the schematic representation of the preprocessing we used: grey blocks represent the offline phase while the yellow blocks represent the online phase (*Source* Author)

9.3 Experiments

We ran training and tests on a machine equipped with Intel(R) Xeon(R) CPU @ 2.30 GHz, 12 GB RAM, and a GPU NVidia Tesla K80. After the outliers removal procedure, we retained from the training set 54554 samples.

In order to present the performances of the models we developed, we use as metrics: the overall accuracy; the true positive rate (TPR); the true negative rate (TNR); the F1 score, which is basically defined as the harmonic mean between precision and recall; and the Cohen's K value, that is a well-known to measure of interrater reliability .

The resulting training set was used to train the three configurations we described in the previous section: the plain convolutional neural network (**KODE-CNN**); the combined use of the previous network with the feed-forward neural network for metadata preprocessed with PCA (**KODE-META-PCA**); finally the same architecture KODE-META-PCA where features of metadata were selected by using PLS/DA (**KODE-META-PLS**).

Moreover, since the PLS/DA during the feature selection process produces also a classification of metadata in severe/non severe defects, in the performance comparison between models we used this as a baseline for a classifier that uses only metadata.

In the Fig. 9.5 we compare performances on both validation and test set. As we can see, the introduction of metadata within the classification process improves significantly the overall accuracy on the task. In particular we reach the 88% of accuracy

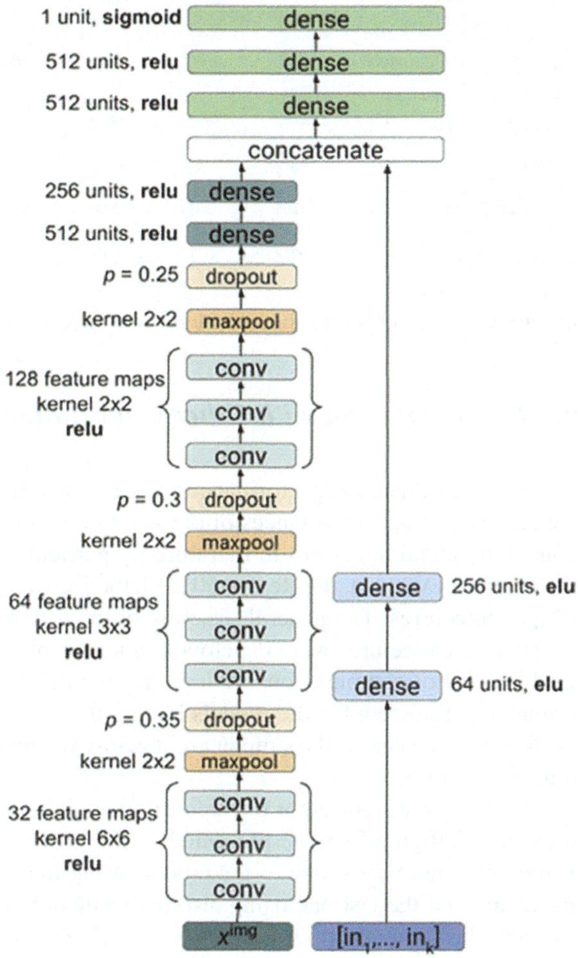

Fig. 9.4 Combined networks architecture used for classification (*Source* Author)

using the KODE+META+PCA model resulting in a quite balanced classification between true positive and true negative samples. With KODE+META+PLS we get almost the 97% of accuracy on the true positives with a bit lower accuracy for true negatives. However, this is not a major problem for our task, as in this context it is more acceptable to misclassify non-severe defects as severe ones than it is to not detect a severe defect when there is one.

	Validation Set			Test Set				
	Acc	*TPR*	*TNR*	*Acc*	*TPR*	*TNR*	*F1 Score*	*Cohen's Kappa*
PLS/DA	.8316	.8380	.8253	.8264	.8269	.8260	.8326	.5000
KODE-CNN	.8186	.9319	.7052	.8165	.9306	.7025	.8370	.6330
KODE-META-PCA	**.8829**	.9205	**.8454**	**.8809**	.9141	**.8478**	**.8874**	**.7618**
KODE-META-PLS	.8715	**.9712**	.7720	.8684	**.9678**	.7689	.8832	.7367

Fig. 9.5 Summary of performances resulting from our experiments (*Source* Author)

9.3.1 Comparison with Other Convolutional Architectures

In order to evaluate the performances of our convolutional architecture as final step we compared the performances of our solution using some of the most well-known convolutional architectures in literature. In particular, as shown in Fig. 9.6 we used the Inception V3 [17], the ResNet50 [18], the DenseNet201 [19], and the MobileNet [20] architectures. The general structure of the new networks is shown in Fig. 9.4.

These architectures were developed to face general tasks and they are characterized by a lot of complex convolutional layers: this has an impact on the number of trainable parameters for the models having these architectures, between 20 and 25 million, with respect to the simpler architecture we developed, that has approximately 6 million parameters.

In this way we got eight different models to be trained from scratch, using the different configurations we presented in the previous section: a family of models using only the images, one of the models using images with the PCA-preprocessed metadata, and the last one using also the results obtained with the PLS to filter out the not relevant metadata's features for the classification.

We did not use the transfer learning because of the nature of our task: in fact, the features learned from the principally available weights, e.g. the ones created using the imagenet dataset, refer to images very different from ours which are, on the other hand, really specific to an industrial context. This choice led us to a slower training process, but in this way, we obtain feature-recognition filters more specific for the shapes of the nonwovens defects.

In Fig. 9.7 we show a summary of the performances for the models using different convolutional architectures. We concentrate on accuracy and on true positive/true negative rates, which are the indexes of interest to evaluate our defect-recognition task. In addition, we report also the occupation size, in megabytes, for all the trained models. The model with the convolutional architecture we developed and described in Fig. 9.4 is underperforming if we consider the models working only on images (Fig. 9.7a). When we introduce in the process also the metadata, we obtain performances that are comparable, and in some cases best than, to the ones reached

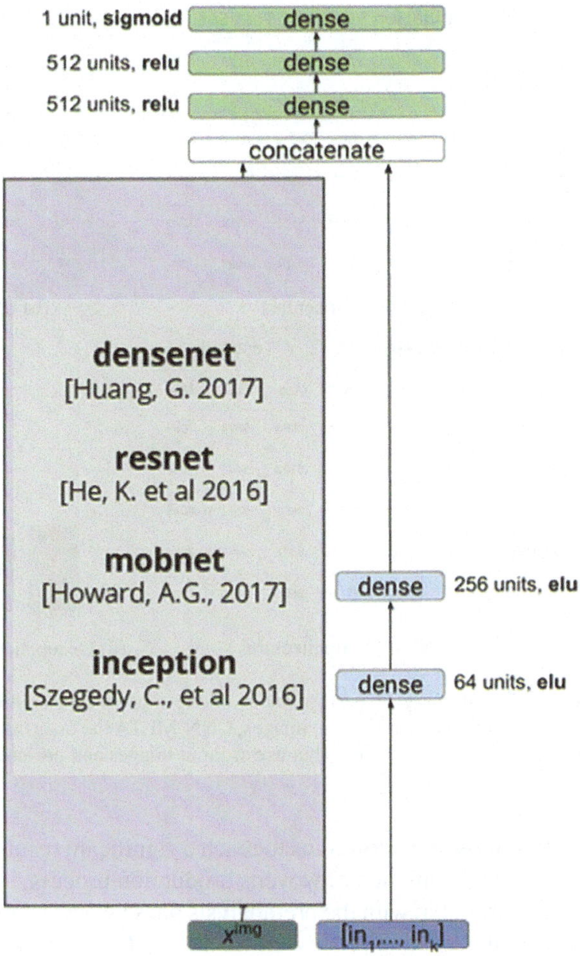

Fig. 9.6 The general structure of the networks used to compare the performances of the one shown in Fig. 9.4 (*Source* Author)

by the models which use more complexes convolutional architectures, as shown in Fig. 9.7b, c.

Moreover, the models using our convolutional architecture is the one with smaller dimension in terms of space occupation: this is an advantage for a model that has to be installed on industrial machines working on a production line.

9.4 Conclusions

In this work we presented an approach to defect classification using a combination of statistical, machine learning, and deep learning techniques using information coming from both images and related metadata. The metadata preprocessing and the archi-

	Validation Set			Test Set		
	Acc	*TPR*	*TNR*	*Acc*	*TPR*	*TNR*
KODE	.8186	.9319	.7052	.8165	.9306	.7025
INCEPTION	**.8595**	.9084	.8105	.8538	.9074	**.8001**
DENSENET	.8590	.9322	**.8407**	**.8551**	.9265	.7837
RESNET	.8546	.9064	.8028	.8523	.9118	.7928
MOBNET	.8497	**.9685**	.7311	.8487	**.9691**	.7284

(a) CNN architectures

	Validation Set			Test Set		
	Acc	*TPR*	*TNR*	*Acc*	*TPR*	*TNR*
KODE	.8829	.9205	.8454	.8809	.9141	**.8478**
INCEPTION	**.8881**	.9289	**.8474**	.8801	.9239	.8364
DENSENET	.8843	.9322	.8364	**.8826**	.9312	.8340
RESNET	.8829	.9252	.8407	.8772	.9218	.8327
MOBNET	.8729	**.9332**	.8125	.8707	**.9322**	.8092

(b) CNN-META architectures

	Validation Set			Test Set		
	Acc	*TPR*	*TNR*	*Acc*	*TPR*	*TNR*
KODE	.8715	**.9712**	.7720	.8684	**.9678**	.7689
INCEPTION	**.8741**	.9460	.8021	**.8702**	.9440	.7964
DENSENET	.8668	.9272	**.8065**	.8658	.9282	**.8035**
RESNET	.8650	.9430	.7871	.8642	.9450	.7834
MOBNET	.8632	.9450	.7814	.8626	.9450	.7807

(c) CNN-PLS architectures

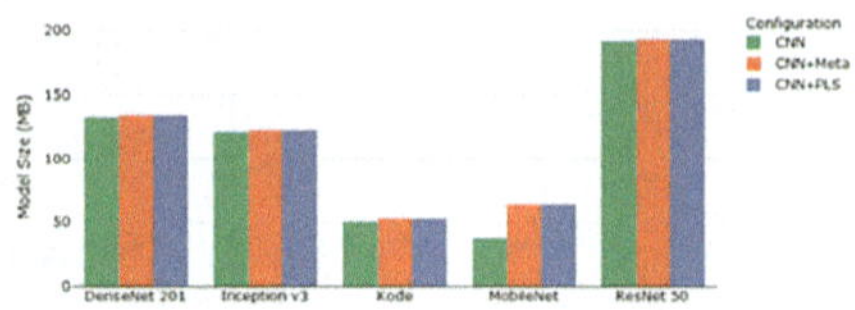

(d) Occupation in MB of the developed models

Fig. 9.7 Summary of performances of the models and architectures under examination. CNN refers to architectures that use images, CNN-META the ones that use as input images and metadata, CNN-PLS the architectures that use as input images and pre-processed metadata (*Source* Author)

tecture we setup allowed us to reach a significant result, identifying the severe defect outcoming from the nonwovens production process.

In particular, with the pre-analysis phase we were able to prepare a cleaner dataset to feed the training process of the network we developed: this allowed us to create architectures using few convolutional layers and a low number of trainable parameters.

Finally we shown as, using our approach to combine the analysis of images and metadata, we can get classification performances which are reached with complex architectures also using a simpler convolutional architecture, gaining a saving in terms of disk occupation as well.

As future improvement, we will work to add to the model the ability to detect the defect into the image using techniques of object detection, e.g. based on YOLO [21, 22] or Faster R-CNN [23] architectures. Another interesting direction to consider is how to embed directly into the in-use vision system the architectures we developed, avoiding to rely on an external server to get the analysis done.

References

1. Kumar, A.: Computer-vision-based fabric defect detection: a survey. IEEE Trans. Ind. Electron. **55**(1), 348–363 (2008)
2. Kunttu, I., Lepisto, L., Rauhamaa, J., Visa, A.J.: Classification method for defect images based on association and clustering. In: Data Mining and Knowledge Discovery: Theory, Tools, and Technology V, International Society for Optics and Photonics, vol. 5098, pp. 19–28 (2003)
3. Gu, I.Y.H., Andersson, H., Vicen, R.: Wood defect classification based on image analysis and support vector machines. Wood Sci. Technol. **44**(4), 693–704 (2010)
4. Krizhevsky, A., Sutskever, I., Hinton, G.E.: ImageNet classification with deep convolutional neural networks. In: Advances in Neural Information Processing Systems, pp. 1097–1105 (2012)
5. Tao, X., Wang, Z., Zhang, Z., Zhang, D., Xu, D., Gong, X., Zhang, L.: Wire defect recognition of spring-wire socket using multitask convolutional neural networks. In: IEEE Transactions on Components, Packaging and Manufacturing Technology, pp. 1–10. IEEE (2018)
6. Essid, O., Laga, H., Samir, C.: Automatic detection and classification of manufacturing defects in metal boxes using deep neural networks. Plos One **13**(11) (2018)
7. LeCun, Y., Bengio, Y., Hinton, G.: Deep learning. Nature **521**(7553), 436 (2015)
8. Soukup, D., Huber-Mörk, R.: Convolutional neural networks for steel surface defect detection from photometric stereo images. In: International Symposium on Visual Computing, pp. 668–677. Springer (2014)
9. Masci, J., Meier, U., Ciresan, D., Schmidhuber, J., Fricout, G.: Steel defect classification with max-pooling convolutional neural networks. In: The 2012 International Joint Conference on Neural Networks, pp. 1–6. IEEE (2012)
10. Cha, Y.J., Choi, W., Büyüköztürk, O.: Deep learning-based crack damage detection using convolutional neural networks. Comput.-Aided Civ. Infrastruct. Eng. **32**(5), 361–378 (2017)
11. Faghih-Roohi, S., Hajizadeh, S., Núñez, A., Babuska, R., De Schutter, B.: Deep convolutional neural networks for detection of rail surface defects. In: 2016 International Joint Conference on Neural Networks, pp. 2584–2589. IEEE (2016)
12. Mujica, L.E., Rodellar, J., Fernandez, A., Güemes, A.: Q-statistic and T2-statistic PCA-based measures for damage assessment in structures. Struct. Health Monit. **10**(5), 539–553 (2011)
13. Chong, I.G., Jun, C.H.: Performance of some variable selection methods when multicollinearity is present. Chemom. Intell. Lab. Syst. **78**(1–2), 103–112 (2005)
14. Favilla, S., Durante, C., Vigni, M., Cocchi, M.: Assessing feature relevance in NPLS models by VIP. Chemom. Intell. Lab. Syst. **129**, 76–86 (2013)
15. Gromski, P.S., Xu, Y., Correa, E., Ellis, D.I., Turner, M.L., Goodacre, R.: A comparative investigation of modern feature selection and classification approaches for the analysis of mass spectrometry data. Analytica Chimica Acta, Elsevier **829**, 1–8 (2014)
16. Galindo-Prieto, B., Eriksson, L., Trygg, J.: Variable influence on projection (VIP) for orthogonal projections to latent structures (OPLS). J. Chemom. **28**(8), 623–632 (2014)
17. Szegedy, C., Vanhoucke, V., Ioffe, S., Shlens, J., Wojna, Z.: Rethinking the inception architecture for computer vision. In: Proceedings of the IEEE Conference on Computer Vision and Pattern Recognition, pp. 2818–2826 (2016)
18. He, K., Zhang, X., Ren, S., Sun, J.: Deep residual learning for image recognition. In: Proceedings of the IEEE Conference on Computer Vision and Pattern Recognition, pp. 770–778. IEEE (2016)
19. Huang, G., Liu, Z., Van Der Maaten, L., Weinberger, K.Q.: Densely connected convolutional networks. In: Proceedings of the IEEE Conference on Computer Vision and Pattern Recognition, pp. 4700–4708 (2017)
20. Howard, A.G., Zhu, M., Chen, B., Kalenichenko, D., Wang, W., Weyand, T., Andreetto, M., Adam, H.: MobileNets: efficient convolutional neural networks for mobile vision applications (2017). arXiv:1704.04861

21. Redmon, J., Divvala, S., Girshick, R., Farhadi, A.: You only look once: unified, real-time object detection. In: 2016 IEEE Conference on Computer Vision and Pattern Recognition, NV, pp. 779–788 (2016). https://doi.org/10.1109/CVPR.2016.91
22. Redmon, J., Farhadi, A.: YOLOv3: an incremental improvement (2018). arXiv:1804.02767
23. Ren, S., He, K., Girshick, R., Sun, J.: Faster R-CNN: towards real-time object detection with region proposal networks. IEEE Trans. Pattern Anal. Mach. Intell. **39**(6), 1137–1149 (2017)

Chapter 10
MIRAI: A Modifiable, Interpretable, and Rational AI Decision Support System

Natalia Mallia, Alexiei Dingli, and Foaad Haddod

Abstract With the recent advancements and results obtained by Deep Learning, several corporations are eager to begin adapting AI into their workflow to benefit from these systems, especially with the emergence of Industry 4.0. However, decision makers find themselves unable to fully trust AI systems from evaluation metrics alone and require some more transparency in their systems. As such, research has gone in the direction of Explainable AI (XAI), where the inner mechanics and reasoning processes of these learning systems are presented in such a way that it is interpretable by humans, opening the Black Box of opaque AI algorithms. This study intends to develop a Big Data and Online Learning-based Explaining AI architecture which utilizes a novel approach of combining Rule-Based Reasoning methodology alongside an explainable Deep Learning framework in order to create a system capable of a higher level of reasoning, or 'Deep Understanding'. Evaluation for this system is to take place by means of standard Machine Learning evaluation metrics such as F1-Scores, Precision, Recall, and ROC Curves. Alongside this, user-feedback evaluation forms from relevant personnel will be used to verify how useful the explanations are in production.

10.1 Introduction

Contemporary Machine Learning (ML) solutions have taken excellent strides in research domains such as Image Recognition, Natural Language Processing, and Speech Automation [2, 4]. However, despite the emergence of more powerful algorithms, they are often considered to be a black box [28]. In real world applications,

N. Mallia (✉) · A. Dingli · F. Haddod
University of Malta, Msida, Malta
e-mail: natalia.mallia.15@um.edu.mt

A. Dingli
e-mail: alexiei.dingli@um.edu.mt

F. Haddod
e-mail: foaad.haddod.15@um.edu.mt

A. Dingli et al. (eds.), *Artificial Intelligence in Industry 4.0*,
Studies in Computational Intelligence 928,
https://doi.org/10.1007/978-3-030-61045-6_10

decision-makers find it difficult to discern whether these systems are effective or not. More importantly, they question if they are able to trust these systems [1]. With the emergence of Industry 4.0, where AI is being implemented in industry to automate process flows in several areas, this becomes a crucial issue that requires due addressing as it greatly affects production. Research efforts have turned towards XAI in order to address the aforementioned concerns. Understanding the processes undertaken by these systems also presents us with the opportunity to develop more cognitive AI systems. As such, we want to develop an intelligent and explainable decision-support system specifically for real-world Industry 4.0 applications. It should be able to completely explain its internal reasoning in such simple terms that a layman will be able to understand the output. Furthermore, we want to allow for improvement in the event that erroneous learning has been identified within the system without disrupting any other learning which took place.

As such, we proposed a **Modifiable, Interpretable, and Rational AI decision support system (MIRAI)**: a novel architecture for an Intelligent Decision Support System (IDSS). The AI component for our IDSS consists of a *Deep Learning* algorithm to obtain inferences from patterns found in data, and a *Reasoning* component to create logical deductions from the data. The combination of inferences and reasoning helps create a deeper, more thorough understanding of the context it is applied in as we can view any patterns present in data from two distinct perspectives. To prove this concept, we used the use-case of a Semiconductor Manufacturing company to develop an Industry 4.0 IDSS solution. Our system assists in reliably predicting downtime in equipment through the use of data from **IoT** sensors. We correlated this data to errors provided by equipment during production and then used ML Techniques to determine patterns and trends which relate to misalignment, and thus, Machine Downtime. Overall, this helps transform the workflow to one which adheres to **Industry 4.0** principles.

10.1.1 Aims and Objectives

This research aims to improve **transparency** within Machine Learning systems and improve performance through XAI. To achieve this, the following objectives were set.

1. Identify the most suitable algorithms for **sequential timestep** data.
2. Extract **explanations** and ensure correct model learning.
3. Use the explanations extracted with Natural Language Generation to create **easy-to-understand** sentences.
4. Iteratively validate the explanations with expert users.

10.2 Related Research

As Ribiero [1] noted, we might be **unaware** of what our Machine Learning models have learned to associate without any form of XAI. Rudin [59] commented that in domains such as medical applications or jurisdiction, these may have **adverse effects** which involve life or death, or imprisoning innocent people. This section covers state-of-the-art methods and tools currently in development to open up these Black Boxes and make AI *Fair, Accountable, and Transparent.*[1]

Types of XAI

Two main schools of thought are at the forefront of XAI; one which is dedicated to **Reverse-Engineering** existing Machine Learning models to derive explanations, whereas the other aims to craft new, **Transparent By Design** algorithms which perform just as well and are inherently explainable.

10.2.1 Reverse-Engineering

Within the context of XAI, Reverse-Engineering methodology attempts to use **statistical** or **transparent AI** techniques to determine the internal workings of the AI model by associating with some form of probability to gauge what the most likely factors to a relevant outcome may be.

Reverse-Engineering techniques may either be **model-agnostic**, that is, they will work with any algorithm, or **model-specific**, where it will only work with a single algorithm.

An example of this is **Feature Selection** [31], where features are removed from data to reduce noise. Either done automatically (for example, removing all low-variance features) or manually (removing all features that do not make sense), this aims to improve the quality of data given to the machine. This, however, is more of a means of shortlisting rather than pinpointing the most relevant features.

One of the most widely-used and highly acclaimed methods, Local Interpretable Model-agnostic Explanations, or more commonly known as **LIME** by Ribiero et al. [1]. It takes each feature independently and assigns a probability for each feature to contribute to the output of the ML model for each data point available, as shown in Fig. 10.1.

Its interpretations tend to be very intuitive to understand and can perform with over 90% fidelity to the processing within the ML model. One of LIME's main weaknesses is that the interpretation is **local**, meaning that it may lose accuracy

[1]https://fatml.org

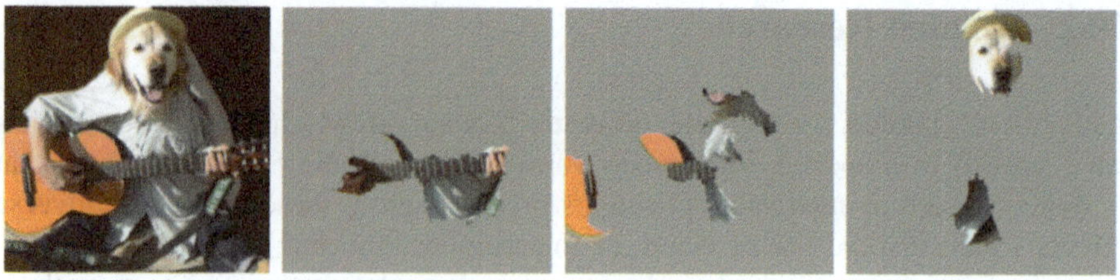

(a) Original Image (b) Explaining *Electric guitar* (c) Explaining *Acoustic guitar* (d) Explaining *Labrador*

Fig. 10.1 An example of an output produced with LIME explaining different local interpretations for classification. *Source* [1]

in its global interpretation, or *the bigger picture*, by narrowing its view to specific instances.

Ribiero et al. [34] further continued their work from LIME and proposed **Anchors**, which works much like LIME, but it provides higher-fidelity predictions. It aims to explain the behaviour of more complicated ML models by deriving important rules they denoted as *Anchors*, which represent adequate conditions for prediction. Much like LIME, however, it shares the bottleneck of being restricted to local explanations, albeit performing slightly better.

One of the very few algorithms which effectively tackles global and local interpretations is LOcal Rule-based Explanations, or **LORE** [35]. It utilizes a genetic algorithm to identify the optimal means of identifying the most important features and creates decision rules based on the results derived. It was able to obtain a score over 90% on **both** local and global explanations, giving it a competitive edge over LIME and Anchors. However, it is also less extensive when presenting its interpretation, which is an area where LIME and Anchors excel in comparison.

10.2.2 Transparent By Design

Contrary to Sect. 10.2.1, this section focuses on algorithms which are created with the intention of having interpretability ingrained into the learning process, thus removing the need to reverse-engineer the system to understand how it works. Earlier models such as Regression [41], Naive Bayes [42], and Decision Trees were considered interpretable as they were mostly associated with statistics rather than AI; as such, they were always fully transparent.

The Accuracy-Interpretability trade-off

The **Accuracy-Interpretability trade-off** [92] states that when improve accuracy, interpretability tends be sacrificed, and vice-versa. Ideally, our algorithms are both accurate and interpretable.

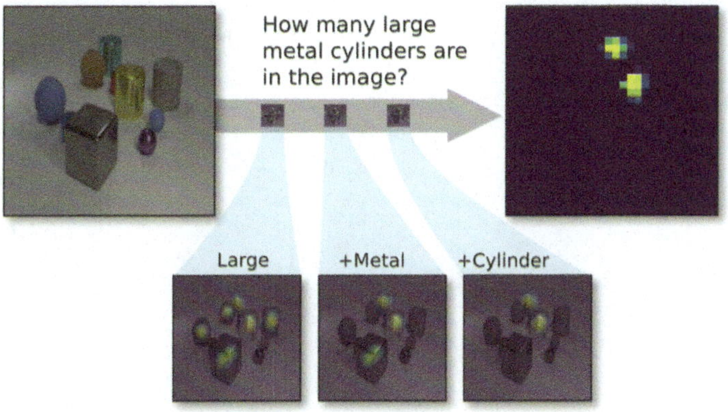

Fig. 10.2 Reasoning performed by TBD-Net. *Source* [50]

In Visual-Task based systems, one algorithm has been crafted to be interpretable by default whilst also boasting fantastic results would be the Transparent-By-Design Network (TBD-Net) by Mascharka et al. [50]. By using attention modules to focus on key areas, they subdivide visual tasks into smaller sub-modules. This technique has proven able to perform visual reasoning on the CLEVR dataset and simultaneously being capable of highlighting the areas which lead to its reasoning task. Figure 10.2 shows an example of such reasoning taking place with TBD-Net. Overall it is robust and performs well, making it a solid approach to solving such problems.

Finally, we look at **Differential Inductive Logic Programming (**δ**ILP)** by Evans et al. [47]. This technique uses Inductive Logic Programming, which consists of a set of pre-defined processing rules used in languages such as DataLog [83], and combines it with Neural Network-like back-propagation on its reasoning to be able to backtrack and learn by itself, given some supplementary rules. It is a remarkably data-efficient technique, and it is also robust to noise in reasoning, making it an excellent algorithm for various applications where more interpretability without sacrificing performance is required. δILP has a baseline of 90% on several performance metrics, reaching state-of-the-art performance compared to other reasoning techniques. It can also be used in conjunction with other networks, such as a Convolutional Neural Network, to perform visual reasoning tasks. It is not compatible with ambiguous data, and is said to be very resource-intensive.

10.2.3 Explaining AI

While XAI is clearly labelled as a crucial factor in an IDSS in Fig. 10.3, few were able to truly incorporate XAI into an IDSS. This is mainly due to the ongoing research in the area, as less interpretable ML models are being used for these assistants [70–72]. As such, we dedicate this section to understanding what is required by XAI from a

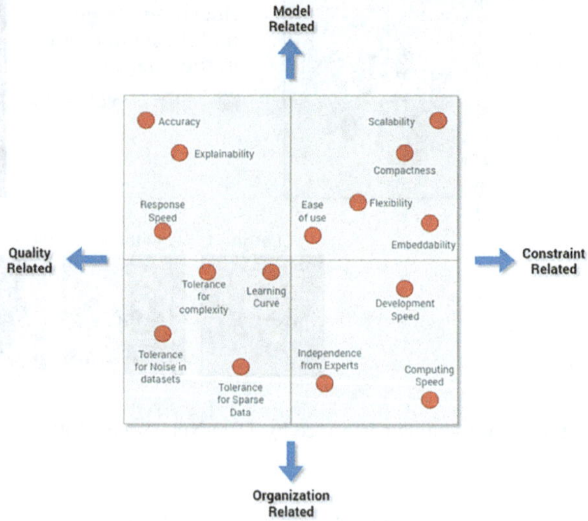

Fig. 10.3 Requirements of an IDSS for management and organizations. Reproduced from [12]

user-perspective. Furthermore, research is currently going into creating either more intuitive interpretations or, more favourably, creating systems where the AI explains itself (**Explaining AI**) [17]. Miller [20] describes human-friendly explanations as an ***"answer to a why-question"***. He proceeds to define what constitutes a good explanation by the following criteria.

- **Contrastive:** Humans are not interested as to why a prediction happened, but rather, why it was made instead of another prediction.
- **Selective:** Usually, listing the entire problem space is ineffective. Instead, users prefer to have one or two reasons for the predictions given.
- **Abnormal-focused:** Any abnormal factors which influence the prediction should be listed as one of the causes.
- **Faithful:** Selection in explanations may poorly convey the truth, and as such, explanations should be as faithful to the true cause.
- **Consistent:** Explanations should be consistent with prior beliefs humans have regarding the domain.
- **General and probable:** A good explanation is one which is easy to generalise. For example, *"Students who study get higher marks"*.

10.3 Methodology

From this research, we have deduced that the best techniques from Model-Agnostic and Reverse-Engineering methods respectively appear to be LIME, Anchors and LORE due to the minimized trade-offs compared to their benefits. On the other hand, from the Explainable by Design section, δILP seems to be the most promising in terms of performance and transparency. As such, we selected these methods for our architecture.

10.3.1 Architecture Overview

Apart from related research, Lippi [86] also released an open challenge for researchers to find a means of combining reasoning techniques with deep learning methodology to create systems which are capable of what they described to be "**Deep Understanding**". That is, creating more cognitive algorithms that not only learn, but also have a more definitive understanding of what they have learned. This concept is based on research carried out by Daniel [85] where he identified two separate systems of thought within the human mind; a fast and intuitive system (*intuition*), and a slow and logical system (*reasoning*). Intuition is responsible for carrying out quick analysis and deductions, an example of which would be identifying correlation. Conversely, reasoning is used for more complex thinking such as identifying causation. We used this knowledge as an opportunity to create an architecture which reflect the requirements of this challenge by proposing a hybrid system. Figure 10.4 demonstrates a high-level overview of this system.

Fig. 10.4 MIRAI system architecture. *Source* Author

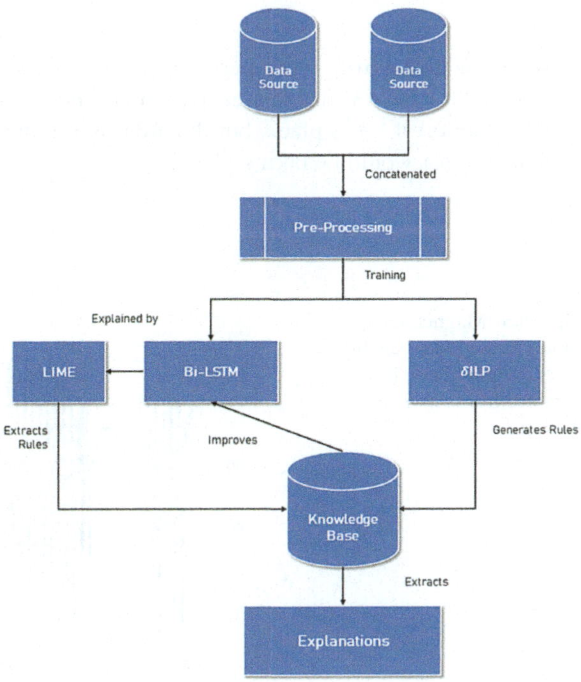

As seen in Fig. 10.4, we obtain and fully clean our data for training and feed it to both a Deep Learning algorithm, in this case a BiLSTM, and a reasoning algorithm, Differential Inductive Logic Programming (δILP). A Knowledge-Base, containing derivations from both algorithms, is subsequently created to hold the information gathered. The Knowledge-Base serves two purposes; to assist the BiLSTM in improving its performance through altering the learned weights to reflect the real world more correctly (This may be done through human intervention or from conflict-detection through rules extracted from δILP). The other purpose is to have the derived results in a standard format which is then used to generate human-readable explanations where the AI may explain itself.

10.3.2 Data Pre-processing

The IoT data used for this study consisted of accelerometer, magnetometer, and gyroscope values representing features for monitoring erroneous behaviour on equipment. Error data was extracted from the associated equipment to determine specific error times. Initially, we ran our analysis on the data to understand what the data looked like before, during, and after an error occurred. From Fig. 10.5, we can see the plot of the magnetometer in the Z-axis in accordance to this time frame. From the error logs, three distinct errors were discovered that correlate perfectly with the three flat sections of the magnetometer plot.

> Knowing that errors were perfectly captured by our IoT sensors led us to pinpoint the most valuable area to predict. The scope was not to target the area where an error takes place, but the area right before it which was to indicate that an error would be taking place.

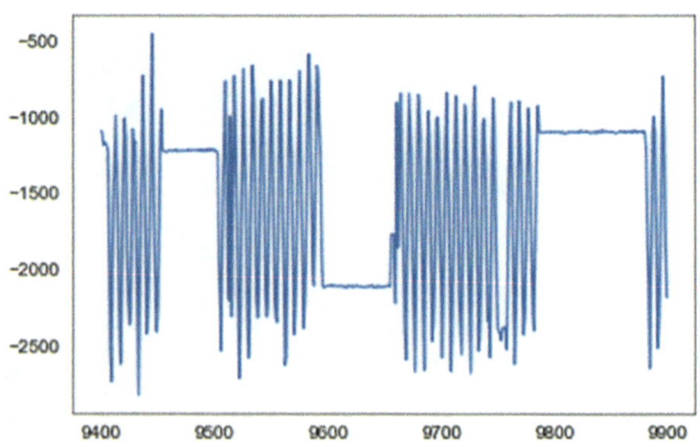

Fig. 10.5 Magnetometer values in the Z-axis against index of time frame it was obtained from. *Source* Author

To accomplish this, we identified the segment between no error (or no relevant error) and an error, and filled the label with gradient values from 0.0 to 1.0, where 0.0 indicates no error and vice-versa. This resulted with the label consisting of values such as 0.2 (not very close to an error) or 0.8 (quite close to an error).

A threshold of 0.75 on the label was set where a flag would be raised to indicate an error. This left the entire upper quartile of values as our main indication just before an error, as it would be close enough to give us more certainty that within this region, an error is close to occurring.

10.3.3 Explanations

Additionally, the extracted rules were further processed to produce natural sentences. Following Miller's concept of constructing good explanations, we shortlisted our explanations to highlight only the most relevant instances to retain simplicity and comprehension. As LIME easily provided those values in its explanations, we formatted the outcome using a custom language template. An example of an explanation would be if the most contributing feature were to be the acceleration in the x-axis, the corresponding explanation would appear as follows.

Explanations

"An error is **75%** likely to occur on Machine **A** at **06.15** a.m. due to the **X-acceleration** having a value **greater than 223**."

10.4 Evaluation

Our main approach to evaluation consists of using performance metrics to measure the effectiveness of the system, and verifying the explanations derived from LIME to further affirm that the BiLSTM shows good learning.

10.4.1 Performance Metrics

To evaluate our performance, we decided to use standard evaluation metrics to determine the ratio of **True Positives (TP)**, **True Negatives (TN)**, **False Positives (FP)**, and **False Negatives(FN)** to determine correctly and incorrectly classified predictions. As such, we selected the **Accuracy**, **Precision**, **Recall**, **F-1 Score**, **ROC**

Table 10.1 Performance results of Normal LSTM versus Convolutional LSTM versus BiLSTM and δILP. *Source* Self

Model	Accuracy	Precision	Recall	F1-Score
LSTM	0.79±(0.017)	0.79±(0.012)	0.80±(0.012)	0.79±(0.012)
Conv. LSTM	0.79±(0.041)	0.79±(0.041)	0.79±(0.045)	0.79±(0.041)
BiLSTM	**0.85±(0.003)**	**0.84±(0.022)**	**0.84±(0.024)**	**0.84±(0.022)**
δILP	**0.99±(0.012)**	**0.99±(0.008)**	**0.99±(0.008)**	**0.99±(0.008)**

Curves, and **Confusion Matrix** as our criteria. Table 10.1 demonstrates the performance of these algorithms.

10.4.2 Explanations

Our approach to validating explanations was to first validate the interpretation for the BiLSTM, as this ensures that we do not have any erroneous learning in our algorithms. Along with machine experts, we then presented our first template for explanations, and **repeatedly refined** the sentence until it resembled the template in Sect. 10.3.3. Furthermore, with their assistance we defined that we know our explanations are relevant if they can highlight any features which are directly related movement in the **X and Y-axis** only, which is where misalignment occurs.

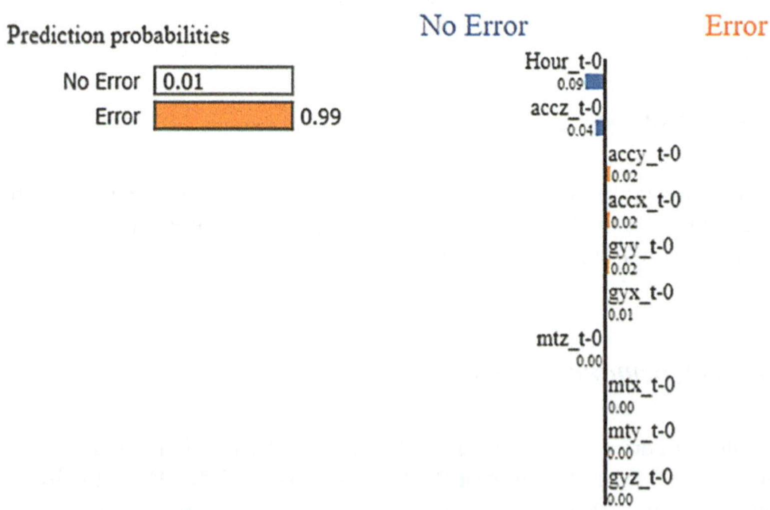

Fig. 10.6 Explanations derived from the BiLSTM via LIME. *Source* Author

Interpretations derived from LIME follow the format of presenting the probability of an explanation of a local point. This must be done for each data point in order to extract a global interpretation. Our main point of interest was to identify which features have been selected as the strongest in a prediction, and if the selected features given the problem space make sense. As seen from Fig. 10.6, The highest contributor to an error is the value of Y-acceleration, closely followed by X-acceleration. However, this model also takes the 'Hour' into consideration which causes some conflict in explanations, as the hour may be the highest contributor on occasion. To avoid this, we drop rules which put too much importance on the hour. Additionally, the highest-contributing features towards error share similar weight, implying that a combination of these values are what contribute to a given prediction rather than a single powerful predictor.

10.5 Conclusion

Following the methodology in Sect. 10.3, we intend to have a system which is both performant and interpretable to further reduce the gap in the accuracy-interpretability trade-off which prevails in Machine Learning and XAI. Apart from this, our research has ventured into the field of Deep Understanding, where we can start to develop system which are more cognitive than typical AI systems through the implementation of similar architectures to MIRAI. The data platform used will also be fully migrated onto a Big Data Online-Learning platform as future work in order to keep up with the ever-growing volume of data created by the use-case. We will eventually connect it with our system to make an automated data system that can stream, learn, and gather relevant explainable analysis in real-time.

References

1. Tulio Ribeiro, M., Singh, S., Guestrin, C.: Why should i trust you?: explaining the predictions of any classifier. In: Proceedings of the 22Nd ACM SIGKDD International Conference on Knowledge Discovery and Data Mining, San Francisco, CA, USA, pp. 1135–1144 (2016)
2. LeCun, Y., Bengio, Y., Hinton, G.: Deep learning. Nature **521**, 436–444 (2015)
3. Rose, L., Fischer, K.: Garbage In, Garbage Out: having useful data is everything. Meas.: Interdiscip. Res. Perspect. **9**(4), 222–226 (2011). https://doi.org/10.1080/15366367.2011.632338
4. Schmidhuber, J.: Deep learning in neural networks: an overview. Neural Netw. **61**, 85–117 (2015). https://doi.org/10.1016/j.neunet.2014.09.003
5. Lake, B., Ullman, T., Tenenbaum, J., Gershman, S.: Building machines that learn and think like people. Behav. Brain Sci. **40**, e253 (2016). https://doi.org/10.1017/s0140525x16001837
6. Botvinick, M., et al.: Building machines that learn and think for themselves: commentary on Lake et al. Behav. Brain Sci. **40**, e255 (2017)
7. Bottou, L.: From machine learning to machine reasoning. Mach. Learn. **94**(2), 133–149 (2014). https://doi.org/10.1007/s10994-013-5335-x
8. Kaplan, J.: Artificial Intelligence: What Everyone Needs to Know, 1st edn. Oxford University Press, New York, N.Y. (2016)

9. Goertzel, B., Pennachin, C.: Artificial General Intelligence. Springer, Berlin Heidelberg (2007)
10. Hudson, D., Manning, C.: Compositional Attention Networks for Machine Reasoning (2018)
11. Schmelzer, R.: From Machine learning to machine reasoning. Medium (2018). https://medium.com/cognilytica/from-machine-learning-to-machine-reasoning-e19fbff154f
12. Dhar, V., Stein, R.: Seven methods for transforming corporate data into business intelligence (1997)
13. Dominic, P. Durai, D., Abdul Aziz, I., Goh, K.N.: A decision support system for performance appraisal. In: Fifth International Conference on Information Technology: New Generations (itng 2008), pp. 899–903 (2008)
14. Dragomir, O., Dragomir, F.: Decision support system integrating fuzzy logic and expert system for optimization of smart grid functioning. In: 2016 International Conference on Control, Decision and Information Technologies (CoDIT), pp. 186–191 (2016)
15. Prakash, N., Sarkar, A.: Development of an intelligent decision support system for an hierarchical business organization. In: 2015 International Conference and Workshop on Computing and Communication (IEMCON), pp. 1–8 (2015)
16. Lou, Y., Caruana, R., Gehrke, J.: Intelligible models for classification and regression. In: Proceedings of the 18th ACM SIGKDD International Conference on Knowledge Discovery and Data Mining, pp. 150–158. ACM, New York (2012)
17. Mittelstadt, B.D., Russell, C., Wachter, S.: Explaining explanations in AI. FAT* 2019 Proceedings (2018). arXiv:abs/1811.01439
18. Fitchett, D., Sondalini, M.: True downtime cost analysis. Feed-Forward (2006)
19. Zhuang, Z., Wilkin, C., Ceglowski, A.: A framework for an intelligent decision support system: a case in pathology test ordering. Decis. Support Syst. **55**(2), 476–487 (2013)
20. Miller, Tim: Explanation in artificial intelligence: insights from the social sciences. Artif. Intell. J. **267**, 1–38 (2019)
21. Doshi-Velez, F., Kim, B.: Towards a rigorous science of interpretable machine learning (2017). arXiv:1702.08608
22. Holsapple, C., Whinston, A.: Business Expert Systems. McGraw-Hill, New York (1987)
23. Sugumaran, V.: Application of agents and intelligent information technologies. IGI Global (2006)
24. George, D., Lehrach, W., Kansky, K., Lázaro-Gredilla, M., Laan, C., Marthi, B., Lou, X., Meng, Z., Liu, Y., Wang, H., others. A generative vision model that trains with high data efficiency and breaks text-based CAPTCHAs. Science 358(6368), eaag2612 (2017)
25. Battaglia, P., Hamrick, J., Bapst, V., Sanchez-Gonzalez, A., Zambaldi, V., Malinowski, M., Tacchetti, A., Raposo, D., Santoro, A., Faulkner, R., others. Relational inductive biases, deep learning, and graph networks (2018). arXiv:1806.01261
26. Fawcett, T.: An introduction to ROC analysis. Pattern Recogn. Lett. **27**(8), 861–874 (2006)
27. Chinchor, N.: MUC-4 evaluation metrics. In: Proceedings of the 4th conference on message understanding, pp. 22–29 (1992). Association for Computational Linguistics, Stroudsburg
28. Lipton, Z.: The mythos of model interpretability. Queue **16**(3), 30:31–30:57 (2018)
29. Guidotti, R., Monreale, A., Ruggieri, S., Turini, F., Giannotti, F., Pedreschi, D. A survey of methods for explaining black box models. ACM Comput. Surv. **51**(5), 93:1–93:42 (2018)
30. Lakkaraju, H., Kamar, E., Caruana, R., Leskovec, J. Interpretable and explorable approximations of black box models (2017). arXiv:1707.01154
31. Guyon, I., Elisseeff, A.: An introduction to variable and feature selection. J. Mach. Learn. Res. **3**, 1157–1182 (2003)
32. Robnik-Sikonja, M., Kononenko, I. Explaining classifications for individual instances. IEEE Trans. Knowl. Data Eng. **20**(5), 589–600 (2008)
33. Baehrens, D., Schroeter, T., Harmeling, S., Kawanabe, M., Hansen, K., Müller, K.R.: How to explain individual classification decisions. J. Mach. Learn. Res. **11**, 1803–1831 (2010)
34. Ribeiro, M., Singh, S., Guestrin, C.: Anchors: high-precision model-agnostic explanations. In: Thirty-second AAAI conference on artificial intelligence (2018)
35. Pastor, E., Baralis, E.: Explaining black box models by means of local rules. In: Proceedings of the 34th ACM/SIGAPP symposium on applied computing, pp. 510–517. ACM, New York (2019)

36. Turner, R.: A model explanation system. In: 2016 IEEE 26th international workshop on machine learning for signal processing (MLSP), pp. 1–6 (2016)
37. Goldstein, A., Kapelner, A., Bleich, J., Pitkin, E.: Peeking inside the black box: visualizing statistical learning with plots of individual conditional expectation. J. Comput. Graph. Stat. **24**(1), 44–65 (2015)
38. Henelius, A., Puolamäki, K., Karlsson, I., Zhao, J., Asker, L., Boström, H., Papapetrou, P.: Goldeneye++: a closer look into the black box. In: International symposium on statistical learning and data sciences, pp. 96–105 (2015)
39. Tolomei, G., Silvestri, F., Haines, A., Lalmas, M.: Interpretable predictions of tree-based ensembles via actionable feature tweaking. In: Proceedings of the 23rd ACM SIGKDD international conference on knowledge discovery and data mining, pp. 465–474. ACM, New York (2017)
40. Kim, B., Wattenberg, M., Gilmer, J., Cai, C., Wexler, J., Viegas, F., Sayres, R.: Interpretability beyond feature attribution: quantitative testing with concept activation vectors (tcav) (2017). arXiv:1711.11279
41. Freedman, D.: Statistical Models: Theory and Practice. Cambridge University Press, Cambridge (2009)
42. Maron, M.: Automatic indexing: an experimental inquiry. J. ACM **8**(3), 404–417 (1961)
43. Wang, F., Rudin, C.: Falling rule lists. In: Artificial intelligence and statistics, pp. 1013–1022 (2015)
44. Chen, C., Rudin, C.: An optimization approach to learning falling rule lists (2017). arXiv:1710.02572
45. Letham, B., Rudin, C., McCormick, T., Madigan, D.: Interpretable classifiers using rules and bayesian analysis: building a better stroke prediction model. Ann. Appl. Stat. **9**(3), 1350–1371 (2015)
46. Lakkaraju, H., Bach, S., Leskovec, J.: Interpretable decision sets: a joint framework for description and prediction. In: Proceedings of the 22nd ACM SIGKDD international conference on knowledge discovery and data mining, pp. 1675–1684 (2016)
47. Evans, R., Grefenstette, E.: Learning explanatory rules from noisy data. J. Artif. Intell. Res. **61**, 1–64 (2018)
48. Pearl, J.: Causality: Models, Reasoning and Inference, vol. 29. Springer, Berlin (2000)
49. Ho, T.: Random decision forests. In: Proceedings of the Third International Conference on Document Analysis and Recognition (Volume 1), pp. 278. IEEE Computer Society, Montreal (1995)
50. Mascharka, D., Tran, P., Soklaski, R., Majumdar, A.: Transparency by design: closing the gap between performance and interpretability in visual reasoning. In: Proceedings of the IEEE conference on computer vision and pattern recognition, pp. 4942–4950 (2018)
51. Braunstein, A., Mézard, M., Zecchina, R.: Survey propagation: an algorithm for satisfiability. Random Struct. Algorithms **27**(2), 201–226 (2005)
52. Hinton, G.: Deep belief networks. Scholarpedia **4**(5), 5947 (2009)
53. Shrikumar, A., Greenside, P., Kundaje, A.: Learning important features through propagating activation differences. In: Proceedings of the 34th International Conference on Machine Learning, vol. 70, pp. 3145–3153 (2017)
54. Krishnan, S., Wu, E.: PALM: machine learning explanations for iterative debugging. In: Proceedings of the 2Nd workshop on human-in-the-loop data analytics, pp. 4:1–4:6. ACM, New York (2017)
55. Higgins, I., Matthey, L., Pal, A., Burgess, C., Glorot, X., Botvinick, M., Mohamed, S., Lerchner, A.: beta-VAE: learning basic visual concepts with a constrained variational framework. ICLR **2**(5), 6 (2017)
56. Adler, P., Falk, C., Friedler, S., Nix, T., Rybeck, G., Scheidegger, C., Smith, B., Venkatasubramanian, S.: Auditing black-box models for indirect influence. Knowl. Inf. Syst. **54**(1), 95–122 (2018)
57. Wang, J., Fujimaki, R., Motohashi, Y. Trading interpretability for accuracy: oblique treed sparse additive models. In: Proceedings of the 21th ACM SIGKDD International Conference on Knowledge Discovery and Data Mining, pp. 1245–1254. ACM, New York (2015)

58. Lee, H., Yune, S., Mansouri, M., Kim, M., Tajmir, S., Guerrier, C., Ebert, S., Pomerantz, S., Romero, J., Kamalian, S.: An explainable deep-learning algorithm for the detection of acute intracranial haemorrhage from small datasets. Nat. Biomed. Eng. **3**(3), 173 (2019)
59. Rudin, C.: Stop explaining black box machine learning models for high stakes decisions and use interpretable models instead. Nat. Mach. Intell. **1**(5), 206–215 (2019)
60. Bradley, A.: The use of the area under the ROC curve in the evaluation of machine learning algorithms. Pattern Recogn. **30**(7), 1145–1159 (1997)
61. Belciug, S., Gorunescu, F. Intelligent Decision Support Systems-a Journey to Smarter Healthcare. Springer Nature, Berlin (2020)
62. Stehman, S.: Selecting and interpreting measures of thematic classification accuracy. Remote Sens. Environ. **62**(1), 77–89 (1997)
63. Kuhn, M., Johnson, K.: Applied Predictive Modeling, vol. 26. Springer, Berlin (2013)
64. Murphy, K.: Machine learning: a probabilistic perspective. MIT Press, Cambridge (2012)
65. Berger, J.: Statistical decision theory and Bayesian analysis. Springer Science and Business Media, New York (2013)
66. DeGroot, M.: Optimal Statistical Decisions, vol. 82. Wiley, Hoboken (2004)
67. Claesen, M., De Moor, B.: Hyperparameter search in machine learning (2015). arXiv:1502.02127
68. Bhagwat, R., Abdolahnejad, M., Moocarme, M.: Applied Deep Learning with Keras: Solve Complex Real-life Problems with the Simplicity of Keras. Packt Publishing, Birmingham (2019)
69. Yerushalmy, J.: Statistical problems in assessing methods of medical diagnosis, with special reference to X-ray techniques. Public Health Rep. **1896–1970**, 1432–1449 (1947)
70. Ahmad, S., Simonovic, S.: An intelligent decision support system for management of floods. Water Resour. Manag. **20**(3), 391–410 (2006)
71. Chan, F., Jiang, B., Tang, N.: The development of intelligent decision support tools to aid the design of flexible manufacturing systems. Int. J. Prod. Econ. **65**(1), 73–84 (2000)
72. Fernandes, K., Vinagre, P., Cortez, P.: A proactive intelligent decision support system for predicting the popularity of online news. In: Portuguese Conference on Artificial Intelligence, pp. 535–546 (2015)
73. Dasgupta, D., Gonzalez, F.: An intelligent decision support system for intrusion detection and response. In: International Workshop on Mathematical Methods, Models, and Architectures for Network Security, pp. 1–14 (2001)
74. Matsatsinis, N., Moraitis, P., Psomatakis, V., Spanoudakis, N.: An agent-based system for products penetration strategy selection. Appl. Artif. Intell. **17**(10), 901–925 (2003)
75. Robert, C.: The Bayesian choice: from decision-theoretic foundations to computational implementation. Springer Science and Business Media, New York (2007)
76. Edwards, L., Veale, M.: Slave to the algorithm: Why a right to an explanation is probably not the remedy you are looking for. Duke Law Tech. Rev. **16**, 18 (2017)
77. Theodorou, A., Wortham, R., Bryson, J.: Designing and implementing transparency for real time inspection of autonomous robots. Connect. Sci. **29**(3), 230–241 (2017)
78. Gunning, D.: Explainable artificial intelligence (xai). Defense Advanced Research Projects Agency (DARPA), nd Web, 2 (2017)
79. Powers, D.: Evaluation: from precision, recall and F-measure to ROC, informedness, markedness and correlation. J. Mach. Learn. Technol. **2**, 37–63 (2011)
80. Bounhas, I., Elayeb, B., Evrard, F., Slimani, Y.: Information reliability evaluation: from Arabic storytelling to computer sciences. J. Comput. Cult. Herit. (JOCCH) **8**(3), 1–33 (2015)
81. Altman, D., Bland, J.: Diagnostic tests. 1: sensitivity and specificity. BMJ: Br. Med. J. **308**(6943), 1552 (1994)
82. Bengio, Y., Boulanger-Lewandowski, N., Pascanu, R.: Advances in optimizing recurrent networks. In: 2013 IEEE International Conference on Acoustics, Speech and Signal Processing, pp. 8624–8628 (2013)
83. Huang, S., Green, T., Loo, B.: Datalog and emerging applications: an interactive tutorial. In: Proceedings of the 2011 ACM SIGMOD International Conference on Management of data, pp. 1213–1216 (2011)

84. Molnar, C.: Interpretable machine learning (2018). https://christophm.github.io/interpretable-ml-book/
85. Daniel, K.: Thinking, fast and slow. Farrar, Straus, and Giroux, New York (2011)
86. Lippi, M.: Reasoning with deep learning: an open challenge. In: 2016 AI*IA Workshop on Deep Understanding and Reasoning: A Challenge for Next-Generation Intelligent Agents, URANIA 2016, vol. 1802, pp. 38–43 (2017)
87. Chawla, N., Bowyer, K., Hall, L., Kegelmeyer, W.: SMOTE: synthetic minority over-sampling technique. J. Artif. Intell. Res. **16**, 321–357 (2002)
88. He, H., Bai, Y., Garcia, E., Li, S.: ADASYN: Adaptive synthetic sampling approach for imbalanced learning. In: 2008 IEEE International Joint Conference on Neural Networks (IEEE World Congress on Computational Intelligence), pp. 1322–1328 (2008)
89. Nguyen, K., Medjaher, K.: A new dynamic predictive maintenance framework using deep learning for failure prognostics. Reliab. Eng. Syst. Saf. **188**, 251–262 (2019)
90. Graves, A., Schmidhuber, J.: Framewise phoneme classification with bidirectional LSTM and other neural network architectures. Neural Netw. **18**(5–6), 602–610 (2005)
91. Zhou, Q., Wu, H.: NLP at IEST 2018: BiLSTM-attention and LSTM-attention via soft voting in emotion classification. In: Proceedings of the 9th Workshop on Computational Approaches to Subjectivity, Sentiment and Social Media Analysis, pp. 189–194 (2018)
92. Gorzalczany, M., Rudzinski, F.: Handling fuzzy systems' accuracy-interpretability trade-off by means of multi-objective evolutionary optimization methods-selected problems, pp. 791–798. Technical Sciences, Bulletin of the Polish Academy of Sciences (2015)

Chapter 11
Multi-Agent Reinforcement Learning for the Energy Optimization of Cyber-Physical Production Systems

Jupiter Bakakeu, Schirin Baer, Hans-Henning Klos, Joern Peschke, Matthias Brossog, and Joerg Franke

Abstract This chapter proposes an artificial intelligence based solution for the efficient operation of a heterogeneous cluster of flexible manufacturing machines with energy generation and storage capabilities in an electricity micro-grid featuring a high volatility of electricity prices. The problem of finding the optimal control policy is first formulated as a game theoretic sequential decision making problem under uncertainty, where at every time step the uncertainty is characterized by future weather dependent energy prices, high demand fluctuation, as well as random unexpected disturbances on the factory floor. Because of the parallel interaction of the machines with the grid, the local viewpoints of an agent are non-stationary and non-Markovian. Therefore, traditional methods such as standard reinforcement learning approaches that learn a specialized policy for a single machine are not applicable. To address this problem, we propose a multi-agent actor-critic method that takes into account the policies of other participants to achieve explicit coordination between a large numbers of actors. We show the strength of our approach in mixed cooperative and competitive scenarios where different production machines were able to discover different coordination strategies in order to increase the energy efficiency of the whole factory floor.

J. Bakakeu (✉) · M. Brossog · J. Franke
Friedrich–Alexander-Universität Erlangen–Nürnberg, Egerlandstr. 7-9, 91058 Erlangen, Germany
e-mail: jupiter.bakakeu@faps.fau.de

M. Brossog
e-mail: matthias.brossog@faps.fau.de

J. Franke
e-mail: joerg.franke@faps.fau.de

S. Baer · H.-H. Klos · J. Peschke
Siemens AG, Gleiwitzer Str. 555, 90475 Nuremberg, Germany
e-mail: schirin.baer@siemens.com

H.-H. Klos
e-mail: hans-henning.klos@siemens.com

J. Peschke
e-mail: joern.peschke@siemens.com

A. Dingli et al. (eds.), *Artificial Intelligence in Industry 4.0*,
Studies in Computational Intelligence 928,
https://doi.org/10.1007/978-3-030-61045-6_11

11.1 Introduction

Today the cost-efficient and stable supply of energy for production processes is a topic of increasing relevance. Changes in the structure of power generation towards decentralized systems and the growing share of renewable energy sources such as photovoltaics and wind power lead to a volatile situation in the supply of electrical energy. The concept in which electricity generation solely follows a demand based on statistical forecasts is no longer applicable. As a consequence, new concepts that harness the flexibility of energy consumption on the consumer side are becoming increasingly important, resulting in new developments in energy markets where a flexible consumer can get incentives for providing flexibility and participating in specific programs.

In this context, Cyber-Physical Production Systems (CPPS) with multiple possible plant topologies and the ability to produce huge number of product variants can make a significant contribution to a transition to a flexible energy demand. However, ensuring such flexibility in the industry 4.0 context requires a sophisticated control strategy to intelligently supervise the power consumption of the underlying components at every given time so that the total energy consumption follows given load profile boundaries. Moreover, the control strategy has to ensure a high productivity while coping with the uncertainties that can arise from the high fluctuations of weather-dependent energy prices, fast market changes as well as random unexpected disturbances in the factory floor.

Traditionally, analytical and simulation-based approaches are used to compute the optimal control strategy for the connected loads, distributed generation and storage components. However, these approaches often require complex models with precise descriptions of system features and functions. The engineering efforts for these systems are often a crucial aspect, which restricts implementations in the large scale. This problem becomes even more important in the industry 4.0 context, where fast market changes can lead to frequent changes in product and process configurations and therefore to the system models.

For these reasons, model-free control solutions and especially new artificial intelligence algorithms are considered as valuable alternatives or supplements [11]. One of the most popular model-free control paradigms that has gained in importance in recent years is Reinforcement Learning (RL) [1, 12]: a model-free algorithm that does not require system identification steps or apriori knowledge. While being very popular in single agent settings (see [2, 4–6]), a naive application of RL in a multi-agent environment (as it is the case for CPPS) is likely to perform poorly. One issue is the non-Markovian nature of the resulting environment: since the action of one agent can affect the outcome of the action of other agents, the single perspective of an agent, which mainly consists of partial observations, becomes non-stationary in the sense that changes that cannot only be explained by the inherent stationary dynamics of the environment can occur. It is therefore important to coordinate the choices of actions of all agents in order to achieve the intended effects. Furthermore, an agent trained in a single environment setting is inherently bounded by the task description.

A slight modification of the task or an evolution of the environment creates new pressures for retraining.

In order to solve these problems, many have worked on unsupervised exploration and skill acquisition methods such as *intrinsic motivation* [3] and *self-play* [8–10]. Rather than training the agents to solve a specific task in a predefined environment, the agent are conditioned in a autocurricula environment whereby the agents have to continuously coordinate their strategies (via competition or cooperation) in order to solve an ever evolving task. As a result, the agents learn an ensemble of emergent skills and coordination strategies rather than a specific complex policy making them more robust to environment changes.

There has been much success in leveraging multi-agent autocurricula to solve multi-agent problems such as [7, 8] both in discrete and continuous real-time domains. In both cases, self-play approaches provided the agents with a perfectly tuned curriculum. The success in these settings inspires confidence that inducing autocurricula in physically grounded and open-ended environments such industrial micro-grids could eventually enable the agents controlling the grid components to acquire an unbounded number of robust and emergent skills that can lead to a global resource optimization. In this chapter, we investigate whether the idea of autocurricula multi-agent reinforcement learning environments can yield fruit in solving the problem of the energy optimization of an heterogeneous cluster of cyber-physical production systems with energy generation and storage capabilities in an electricity micro-grid featuring a high volatility of electricity prices.

In more detail, we first formulate the energy optimization problem as an observable Markov game which we then use to design an autocurricula environment that simulates a micro-grid of flexible manufacturing machines with energy generation and energy storage capabilities. Afterwards, we introduce several multi-agent tasks with comparative goals, where the different agents would need to learn not only highly developed skills but also coordinate the learned strategies in order to achieve the intended global effect. We then train the agents using a distributed implementation of a recent policy gradient algorithm, Proximal Policy Optimization (PPO) [6] in a framework of centralized training with decentralized execution. By adding a simple exploration curriculum to aid exploration in the environment, we found that in both cooperative and competitive scenarios different production machines were able to discover different coordination strategies in other to increase the energy efficiency of the whole factory floor.

The remainder of this chapter is structured as follows. In Sect. 11.2, backgrounds and related work on load management systems as well as multi-agent reinforcement learning algorithms are presented. Then, we formulate the energy optimization problem as a sequential decision making problem under uncertainty in Sect. 11.3. Section 11.4 gives a detailed description of our training algorithm and the autocurricula environment. Experimental results demonstrating the strength of our approach are presented in Sect. 11.5. Finally, we summarize the paper and discuss some directions for future work in Sect. 11.6.

11.2 Backgrounds and Related Work

11.2.1 Load Management Systems

A growing number of works on load management systems have primarily focused on industrial load control and have generally ignored the volatility of the energy prices [11, 14–16]. Most recent works [13, 17–19] took into account this factor and showed that analytical approaches can be used to compute the optimal time to change the energetic state of a machine for energy control purposes without sacrificing the system throughput. Energy-aware scheduling was also considered to optimize the production plan using mixed integer programming [20, 21]. However, all the proposed solutions rely on fixed product and machine configurations and thus are not applicable in a dynamic context. Furthermore, these solutions require a precise estimation of the model of the production machine as well as the associated parameters. Since the acquisition of such models mostly rely on expert knowledge, the proposed solutions are not scalable and not cost effective. In this chapter, we focus on a model-free control solution that use RL and do not require an explicit system identification.

11.2.2 Reinforcement Learning

Reinforcement learning (RL) is one of the main classes of machine learning methods that has been successfully applied to solve sequential decision-making problem under uncertainties [22]. In such a setting, an agent learns to act using a (partial observable) Markov Decision Process (MDP) [1] formalism. At each time step, the agent selects an action from a finite set of possible actions. After executing the selected action, the environment is updated and the agent receives a rewards signal, which can also be interpreted as a punishment from the environment. The goal of the agent is to maximize the discounted reward in the long run. To achieve this, the agent has to compute the optimal policy based on interactive learning with the environment (see Fig. 11.1).

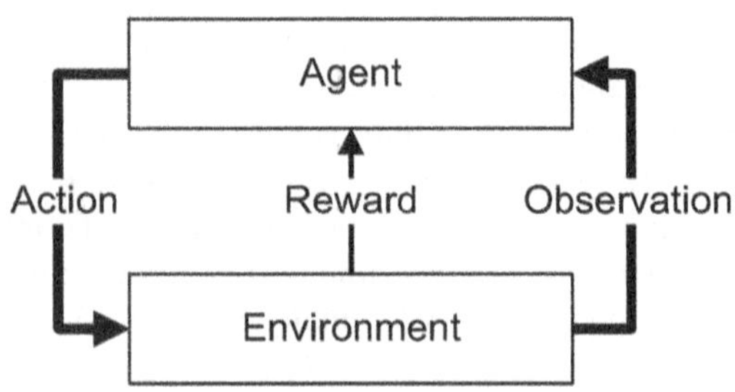

Fig. 11.1 The agent-environment interaction in reinforcement learning [1]

The literature provide many examples of RL algorithms applied to energy management problems. In 1996. Anderson et al. applied Q-learning to modulate the output of the PI controller for a heating coil [23, 24]. In the same direction, Ruelens et al. [25, 26], De Somer et al. [27] and Kazmi et al. [28] were able to apply RL methods to reduce the domestic water heating costs. Jiang et al. implemented a hierarchical multi-agent Q learning approach for dynamic demand response and distributed energy resource management in a micro-grid [29]. In recent studies [30–32], the authors implemented a multi-agent Deep Q-Network (DQN) [2] approach where each agent controlled a house. The agent could consume/store energy, request energy from or grant energy to neighbors, deny energy to neighbors or purchase it from the grid.

11.2.3 Multi-Agent Reinforcement Learning

In Multi-Agent Reinforcement Learning (MARL), multiple agents perform concurrently in a single environment [33–36]. The simplest approach to learning in multi-agent settings is to train each agent independently. This was attempted with Q-learning in [36], but does not perform well in practice [37] since many applications assume interdependency between the agents. More recently, deep neural networks have been successfully used in RL to approximately represent policy and value functions. It is therefore not surprising that deep neural networks have also been applied to MARL settings [38, 39], allowing multi-agent learning in high-dimensional/continuous state spaces. However, applying standard RL methods to MARL problems is challenged by some limitations, such as non-stationary of the environment from the perspective of individual agents [40–42], lack of coordination/communication in cooperative settings [41, 43–45], credit assignment in cooperative settings with global rewards [42, 46, 47], and the failure to take opponent strategies into account when learning agent policies [48].

Most relevant to this work are recent approaches [42, 44] that propose an actor-critic framework consisting of centralized training with decentralized execution as well as some approaches that utilize attention in a fully centralized multi-agent setting [49, 50]. The proposed method uses separate centralized critics for each agent, which take all other agent's actions and observations as input during the training phase, and actors whose policies are conditioned only on local information. This practice reduces the non-stationary of multi-agent environments, as considering the actions of other agents to be part of the environment makes the state transition dynamics stable from the perspective of one agent. In practice, these ideas greatly stabilize learning because of the reduced variance in the value function estimates.

Our proposed solution follows these recent developments and propose a centralized critic for PPO applied in a multi-agent setting. It is more flexible than the aforementioned approaches in the sense that it is able train policies in environments with any reward setup, observation and action spaces making it broadly applicable to different environment types including industrial micro-grid environments.

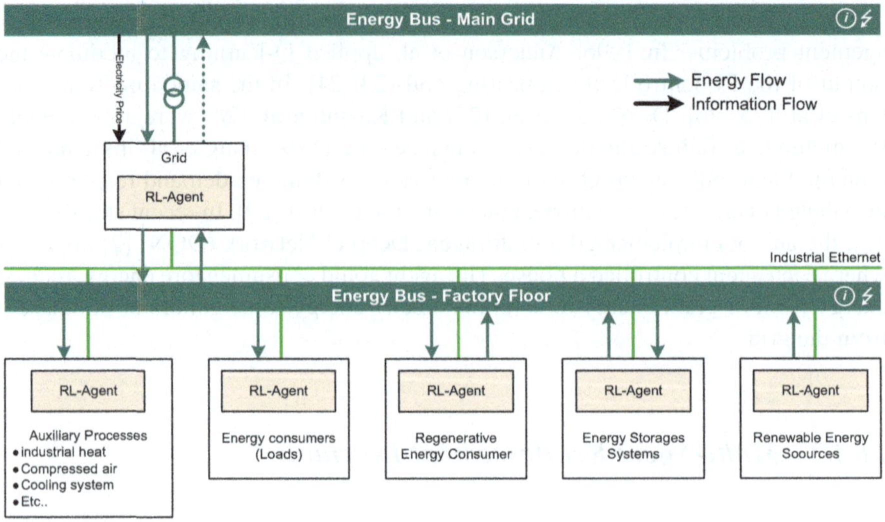

Fig. 11.2 System Overview. Every grid-component is controlled by a reinforcement learning agent. *Source* Author

11.3 System Model and Problem Formulation

The chapter addresses the problem of optimally controlling an industrial micro-grid featuring a large share of renewable energy and a high volatility of electricity prices. We consider a micro-grid as a localized group of energy sources, loads and storage components that can operate in two distinct modes: grid-connected mode and isolated mode. In grid-connected mode, the micro-grid system has the possibility to buy/sell energy from/to the macro-grid in order to meet the load demand. The challenge in connected mode is to reduce the total energy cost. In isolated mode, the micro-grid can function autonomously in the sense that enough energy can be generated in the grid to supply the loads. In this case the challenge is to meet the load demand and to balance the power flow between the components of the grid. In our setting, the grid is executed in mixed mode. In other words, the industrial micro-grid, we are considering, can switch from one mode to another.

11.3.1 System Model

As illustrated in Fig. 11.2, the micro-grid system consists of the following components: renewable energy sources such as photovoltaic systems, electricity loads representing production machines, regenerative energy consumers, energy storages systems such as batteries and auxiliary process loads for air compressors and cooling systems. The components share the same energy bus, enabling a transfer of electrical energy between the components. The proposed system model is equally applicable to ac and dc micro-grids and a simple formal description of the components can be described as follow:

- ***Renewable energy sources*** are renewable energy generators such as photovoltaic systems or wind turbines. The power generated depends on environmental and weather factors such as solar radiation profile and wind speed. It is therefore not predictable. Energy generators are characterized by two operating states: on and off. The generators are considered to be in the off-state when the electrical power generated by the energy sources is negligible for example because of weather conditions such as cloudy weather or nighttime. We assume that if the generators are not in off state, then the power generated is constant and equal to the maximum electricity power that can be produced by the source. Therefore

$$P_G = \begin{cases} 0 & \text{if } state = off \\ P_G max & \text{otherwise} \end{cases} \tag{11.1}$$

- ***Energy storage systems***. In order to take full advantage of renewable energy sources, it is vital to have energy storage systems capable of handling variations in energy production. In our environment, we consider batteries or super-capacitor systems that consume a constant energy when charging and produce a constant energy when discharging. In addition, the operation of energy storage systems has to be carefully designed and controlled to protect them from damages that are: overcharging and overdischarging. Therefore, each storage system is also characterized by a maximum and minimum state-of-charge (SoC). The energy storage systems have three required states: $charging$, $discharging$ and $idle$. The electrical power consumed or released by the component highly depends on the current operational state and the component's state-of-charges.
- ***Energy loads*** represent production machines that consume energy in order to execute a production task. A production machine can have different operation states: $poweredoff$, $executing$, $standby$, etc. Depending on the production task and the operation state, a production machine can follow a predefined or a characteristic load profile. For a fixed set of production tasks, the load profile of a machine can be measured and approximated. In our setting, we consider 3 basic load profile as illustrated in Fig. 11.3. Any machine's load profile can be seen as a linear combination of this basic load profiles. In isolated mode, the micro-grid cannot guarantee continuous power supply to the load because it is often influenced by the unpredictable power generation. When the generated power is not able to drive the loads, the noncritical loads must shut down.
- ***Regenerative energy consumers*** are energy consumers with the exception that these consumers can recuperate energy for a small period of time (less than a minute). In this case we consider the load to be constant and negative. Figure 11.4 shows a typical load profile.
- ***Auxiliary processes*** are energy consumer processes in the factory floor that do not execute a manufacturing task directly, but are required by the production task.

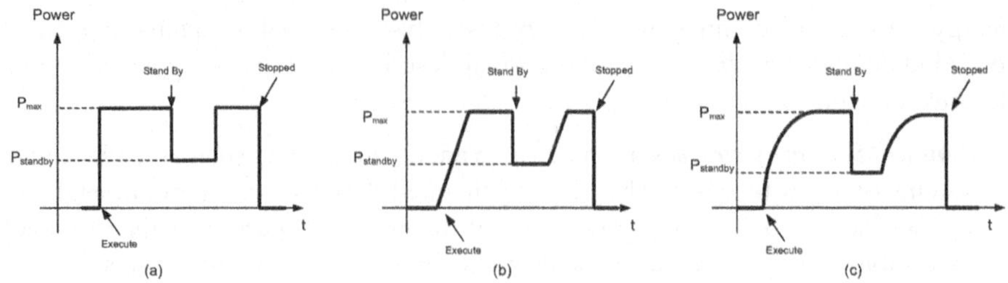

Fig. 11.3 Basic load profiles of the energy loads. In *execute*-state, the electrical power consumed P_i can be constant and equal to P_{max} (a), increase linearly (b) or exponentially (c) to P_{max}. *Source* Author

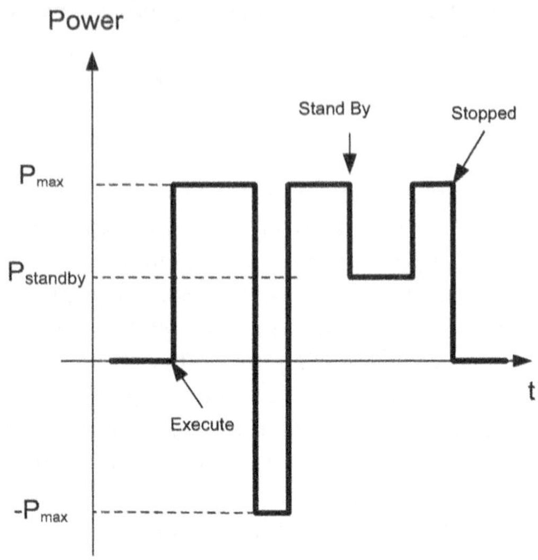

Fig. 11.4 Basic load profiles of regenerative energy consumers. In *execute*-state, the electrical power consumed P_i can be negative for a short period of time. *Source* Author

These are for example compressors or cooling systems. Auxiliary processes have two required states: *off* and *on*.

- The ***grid component*** represents the main grid and is responsible for buying/selling electricity from/to the main grid. It has three state: *buying*, *selling*, *idle*. If the grid agent is in *idle*-state or in *off*-state, the industrial micro-grid can be considered to be executed in isolated mode because there are no interaction with the main grid.

11.3.2 Problem Formulation

The main objective is to minimize the total cost the energy bought from the main grid to achieve a high productivity while considering future energy prices and weather dependent renewable energy generators. Let M denotes the set of components of the micro-grid (or controllable machine components), such that $M_i \in \mathrm{M}, \forall i \in [0, M-1]$ with $M \in \mathbb{N}$. We assume that each component M_i can take an energy state $s_{i,t}$ (i.e. "stopped", "running", "aborted", "standby", etc.) at the time step $t \in \mathbb{N}$ and the dynamic power consumption P_i^t of M_i solely depends on the current state $s_{i,t}$ of the component and the current time step t: $P_i^t = P_i(t, s_{i,t})$. The total energy requested/sold from/to the main grid E_i^t at time interval Δt over the complete time horizon $T \in \mathbb{N}$ is therefore the sum of the power consumed/generated of all components during the time horizon. Please notice that P_i^t can also be negative in the case of renewable energy sources or regenerative energy consumers for example.

$$E_i = \sum_{t=0}^{T-1} P_i^t \cdot \Delta t, \forall i \in [0, M-1] \tag{11.2}$$

If $\lambda_t^- \in \mathbb{R}$ is the actual energy price and $\lambda_t^\sim \in \mathbb{R}$ the forecasted energy price at a time step t, the optimization problem can be formulated for the time horizon T as follow:

$$minimize \sum_{k=0}^{t} \lambda_t^- \cdot \left(\sum_{i=0}^{M-1} P_i(k, s_{i,k}) \cdot \Delta t \right) + \sum_{l=t+1}^{T-1} \lambda_t^\sim \cdot \left(\sum_{i=0}^{M-1} P_i(l, s_{i,l}) \cdot \Delta t \right) \tag{11.3}$$

If $a_l(s_{i,l-1}, s_{i,l})$ denotes the action of changing the state of the component M_i at time l from the state $s_{(i,l-1)}$ to the state $s_{(i,l)}$, the Eq. 11.3 can be rewritten as follow:

$$minimize \sum_{k=0}^{t} \lambda_t^- \cdot \left(\sum_{i=0}^{M-1} P_i(k, s_{i,k}) \cdot \Delta t \right) + \sum_{l=t+1}^{T-1} \lambda_t^\sim \cdot \left(\sum_{i=0}^{M-1} P_i(l, a_l(s_{i,l-1}, s_{i,l})) \cdot \Delta t \right) \tag{11.4}$$

Several constraints should be taken into consideration. One of them is the power balance between the energy demand of the micro-grid and the energy supply from the main grid:

$$\sum_{i=0}^{M-1} E_i \leq E_{max}^\sim, \forall i \in [0, M-1] \tag{11.5}$$

where E_i denotes the total energy demand of the micro-grid component M_i over the time horizon T and $E_{max}^\sim$ the total available energy from the main grid.

In addition, the total energy consumption of the micro-grid at each time interval t should satisfy the upper and lower bound given by the overall load profile and a tolerance interval. This constraint can be expressed as follows:

$$L^{\sim}_{target} - \Delta L^{\sim} \leq \sum_{i=0}^{M-1} P_i(t, s_{i,t}) \leq L^{\sim}_{target} + \Delta L^{\sim} \quad (11.6)$$

with $L^{\sim}_{target}$ being the target load at time step t, $\Delta L^{\sim}$ a symmetric tolerance around $L^{\sim}_{target}$ and $i \in [0, M-1]$

As expressed above, solving the optimization problem is equivalent to compute at each time step the optimal state (by choosing the action $a_l(s_{i,l-1}, s_{i,l})$) so that the resulting total energy cost of the micro-grid for a complete time horizon is globally minimized. If every component of the micro-grid is controlled by an agent, then the optimization problem is equivalent to finding a coordinated strategy for all the agents. In this case, the main challenge lies in the volatility of future energy prices and the direct dependence of future energy consumption from actual decisions. Furthermore, other constraints such as the the throughput, production time, the product quality, the production cost, etc. have to be considered.

11.3.3 Markov Game Formulation

The formulated optimization problem can be transformed into a reinforcement learning task, where each agent controlling a component of the grid has to learn a policy that maximizes a cumulative reward signal derived from the objective function and is conditioned by the constraints formulated in Eqs. 11.5 and 11.6.

If we consider the global state of the environment as the aggregation of the observations of all agents, then the observed state of the environment solely depends on the actions of all the agents and the previously observed global state of the environment. In other words, no external factors besides the actions of all the agents will affect the dynamics of the environment. Under this assumption, the multi-agent reinforcement-learning task satisfies the Markov property and can be therefore formulated as a Markov decision process (MDP).

In this work, we consider a multi-agent extension of Markov decision processes called observable Markov games [33]. A Markov game for N agents is defined by a set of states S describing the possible configurations of all agents, a set of actions $\mathrm{A}_1, \mathrm{A}_2, \ldots, \mathrm{A}_N$ and a set of observations $\mathrm{O}_1, \mathrm{O}_2, \ldots, \mathrm{O}_N$ for each agent. To choose actions, each agent $i\,(i = 1, 2, \ldots, N)$ uses a stochastic policy $\pi_{\theta_i} : \mathrm{O}_i \times \mathrm{A}_i \mapsto [0; 1]$, where θ_i are the parameters of the policy. The policy produces the next state according to the state transition function $T : \mathrm{S} \times \mathrm{A}_1 \times ... \times \mathrm{A}_N \mapsto \mathrm{S}$. Each agent i obtains rewards as a function of the state and agent's action $r_i : \mathrm{S} \times \mathrm{A}_i \mapsto \mathbb{R}$, and receives a private observation correlated with the environment state $o_i : \mathrm{S} \mapsto \mathrm{O}_i$. The initial states are determined by a distribution : $\mathrm{S} \mapsto [0; 1]$. Each agent i aims to maximize its own total expected return $R_i = \sum_{t=0}^{T} \gamma^t \cdot r_i^t$ where γ^t is a discount factor at time step t and T is the time horizon. The *action-value function* is defined as $\mathrm{Q}^{\pi_i}(s_{i,t}, a_t^i) = \mathbb{E}[\mathrm{R}_i^t | s_{i,t}, a_t^i]$, while the *state-value function* is defined as $\mathrm{V}^{\pi_i}(s_{i,t}) = \mathbb{E}[\mathrm{R}_i | s_{i,t}]$. The *advantage function* $\mathrm{A}^{\pi_i}(s_{i,t}, a_t^i) = \mathrm{Q}^{\pi_i}(s_{i,t}, a_t^i) - \mathrm{V}^{\pi_i}(s_{i,t})$ describes

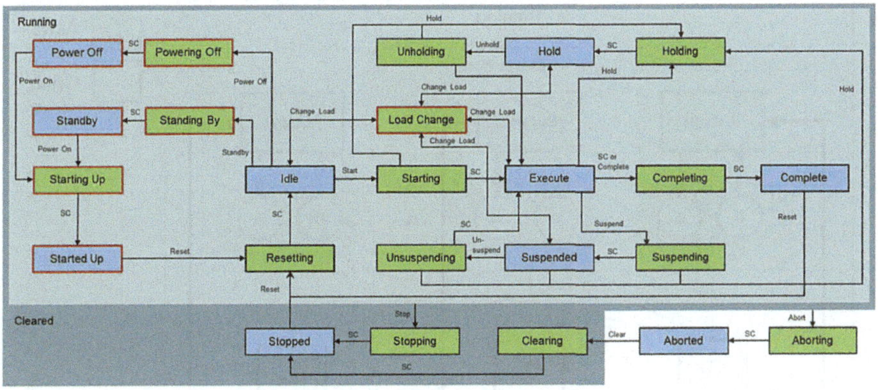

Fig. 11.5 Uniform state representation of a grid component. Every component can be stopped, halted, suspended, powered up or aborted. Once the component reach the *execute*-state then it begins to execute a production task, to generate energy or to store/release energy. *Source* Author

whether taking action a_t^i is better or worse for agent i when in state $s_{i,t}$ than the average action of policy π_{θ_i}.

For our micro-grid, we provide a uniform representation of the operational state of the grid components which is illustrated in Fig. 11.5. The state space of every agent is represented by the operation state of the grid component it controls aggregated with the target load of the grid, the current time step within the production shift, the amount of production tasks to execute as well as the current energy price. Notice that we do not assume inter-dependencies between the production tasks. The action space of each agent is represented by all the actions that can change the operational state of a grid component (See Fig. 11.5). The reward of each agent depends on the type of the grid component it controls and therefore is use case dependent. See Eq. 11.7 in Sect. 11.4.2 for more details.

11.4 Approach

11.4.1 Multi-Agent Proximal Policy Optimization

Since the naive application of standard reinforcement learning algorithms performs poorly in multi-agent settings, our goal is to derive an algorithm that can operate under the following constraints:

- The learned policies can only use local information (i.e. own observations) at execution time.
- There is no particular communication method between the agents, since it can restrict the scalability of the solution.

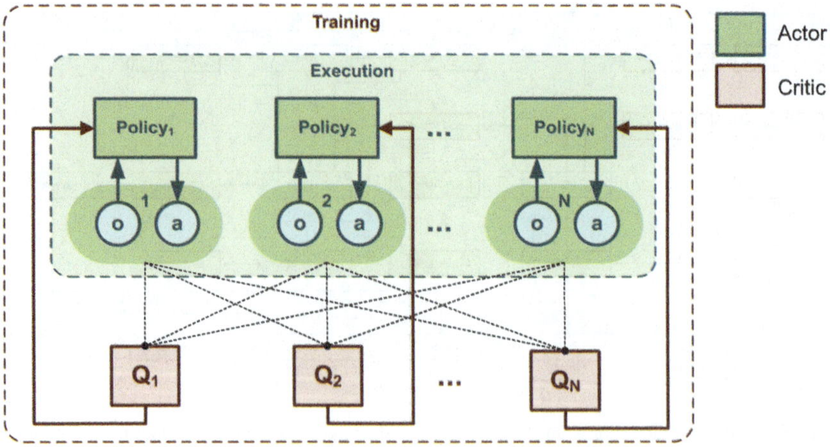

Fig. 11.6 Overview of our multi-agent decentralized actor, centralized critic approach. The centralized action-value function (critic) for each agent takes as input the actions and observations of all agents and predicts the discounted future returns (Q-values) for the agent while the actor produces an action distribution based solely on the agent's observations and actions. *Source* Author

- We do not assume a differential model of the environment dynamics.

Fulfilling the above desiderata would provide a general-purpose multi-agent learning algorithm that could be applied not just to cooperative games with explicit communication channels, but also to competitive games and games involving only physical interactions between agents.

We accomplish our goal by adopting the framework of centralized training with decentralized execution in which agent policies are composed of two separate networks with different parameters: a policy network (actor) that generates an action distribution and a critic network (critic) that predicts discounted future returns. The critic uses extra information to ease training process whilst actors take actions based solely on their own local observations. This is still a valid approach as long as the information used during the training of the agents are not used at test/execution time.

As illustrated in Fig. 11.6, our centralized action-value function (critic) for each agent takes as input the actions and observations of all agents and predicts the discounted future returns for the agent. Additional information can also be included in the inputs. Since each action-value function is learned separately, each agent can have arbitrary rewards including conflicting rewards in competitive scenarios.

Furthermore, we extend the idea of [6] by adopting the proximal policy optimization algorithm (PPO) as the main policy optimization algorithm instead of deep deterministic policy gradient (DDPG) as presented in the paper. PPO is a family of policy optimization methods that use multiple epochs of stochastic gradient ascent to perform each policy update. It performs comparably to or better than DDPG while being much simpler to implement and tune.

11.4.2 Cooperation and Competition

As we seek the emergence of coordination strategies between the components of the micro-grid, the agents controlling the micro-grid component s are trained using *self-play* in an *autocurricula* configuration. This acts as a natural curriculum as these agents are always in a mixed cooperative and competitive setting. For example, the energy loads (i.e. production machines) of the micro-grid are always in competitive settings. If there is not enough electrical energy available in the grids, then some production machines have to turn themselves off so that the production can be carried on by the other machines. In the same direction, cooperation can take place if, for example, the energy storage components need to be switched to the discharging state in order to increase the electrical energy available on the energy bus and thus enable the production machines to perform additional production tasks.

Adopting such a configuration also offers many advantages. If a new successful strategy is adopted by an agent, it implicitly changes the task distribution for the competing agents which are then forced to develop a better strategy. These evolutionary arms races create an implicit autocurriculum in which competing agents constantly create new tasks for each other and thus discover new powerful and robust strategies and counter-strategies. In addition, the multi-agent autocurriculum is open-ended. That means that the training will only stabilizes once equilibrium between the different agents is founded which can only be reached if a suitable coordination strategy between the agents has emerged. Due to the exploration factors, many coordination strategies can also emerge at different stages of the training.

In order to create a competitive setting, competing agents are rewarded with competitive rewards. If an agent wins the game, it receives a positive reward and all other competing agents are penalized by an opposite negative reward. If nobody wins the game, then all competing agent are penalized. In a cooperative setting, agents are given a team based reward; meaning that all members of the team receive a positive reward, if a team member wins the game. To confine agent behavior to a reasonable space, we also introduced an environmental-based penalization.

The success of agents in the autocurricula setting requires the agents to occasionally solve the task (win the game) by random actions. The probability of this happening in most games is minuscule as they require as a prerequisite some fundamental skills like the ability to properly control the operational state machine of the grid component . For example the agent must first quickly learn how to bring a production machine in the $execute$-state in order to start a production task or how to turn on/off the storage component without suspending/halting the production. To overcome this problem, we use simple dense rewards at each step so that the agents can first learn basic motor skills that increase the likelihood that random actions of the agent will result in a relatively small positive reward. We refer to this reward as the *exploration reward*. The exploration reward is then gradually annealed to zero, in favor of the *competition/cooperation reward*, to allow the agents to learn coordination strategies for the remaining training time. This is achieved using a linear annealing factor α. We also introduce an *environmental reward* which aims to penalize action sequences leading to dangerous outcomes (for example system configurations violating the constraints formulated in Eqs. 11.5 and 11.6) and a *productivity reward* which encourage production machines to execute production task s more frequently.

So, at time-step t, if the exploration reward is r_{explo}, the competition reward is r_{comp}, the environmental reward r_{env}, the the productivity reward r_{prod} and T is the termination time-step, then the total agent reward is:

$$r_t = \alpha \cdot r_{explo} + (1 - \alpha) \cdot (r_{comp} + r_{env} + r_{prod}) \tag{11.7}$$

With this formulation, the agents are trained initially on the dense reward for only about 10–20% of the training epochs in order to gain some basic skills first.

11.5 Experiments

Our aim is to show that the agents controlling an industrial micro-grid can learn complex behaviors and coordination strategies that can lead to an optimization of the global energy consumption. Therefore, we train the agents in different competitive and cooperative scenarios. In each scenario, the environment consists of multiples agents with discrete action space, discrete time space and mixed observation spaces. Every agent controls a grid component and may take physical actions in the environment by changing the operational state of the corresponding grid component. We do not assume that all agents have identical action and observation spaces, or act according to the same policy π_θ. The environment of each experiment were implemented as an extension of OpenAI environment Gym [51]. We found this framework useful for creating environments involving complex interactions between agents , while keeping the control and perception problems simple, as we are primarily interested in addressing agent strategies.

Unless otherwise specified, the actor and the critic of each agent is parameterized by a two-layer ReLU MLP with respectively 256 units per layer for the actor and 512 units per layer for the critic. In all of our experiments, we use the Adam optimizer with a learning rate of 0.0001 and the discount factor γ is set to be 0.9. The size of the replay buffer is 10240 and the sampling time horizon is 1024. We update the network parameters after every 1024 sampling steps and we use a batch size of 512. We train with 10 random seeds for environments with stark success/fail conditions (*experiment 2* and *experiment 3*) and with only 2 seeds for *experiment 1*. For the PPO-algorithm, we set the clipping parameter ϵ to 0.1, the entropy coefficient to 0.005 and the exploration noise to 0.2.

This section describes in detail the designed experiments as well as the specific training parameters and analyzes various aspects of the learned behaviors. Source codes for the environments as well as learned policy parameters of the agents in all environments are open-sourced and available at [52].

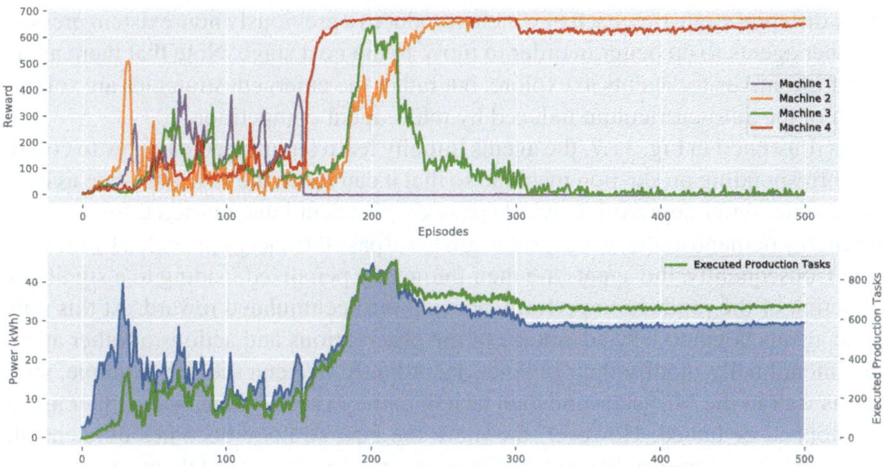

Fig. 11.7 Emergent Skill Progression From Multi-Agent Autocurricula. Through the reward signal agents go through different distinct stages of emergence. **a** Episode 1–30: the different agents learn basic skills (how to execute a production task) and the productivity rise. **b** Episode 30–160: strategy exploitation and competition take place between the agents and stabilize only when *machine 1* adopts a very bad strategy. **c** Episode 160–220: the first coordination strategy emerge. *Machine 2*, *Machine 3* and *Machine 4* start to produce more frequently while *Machine 1* remains in the *idle*-state more often. **d** Episode 220–500: a second coordination strategy emerge. *Machine 2*, *Machine 4* reached their highest productivity while *Machine 3* is frequently halted. *Source* Author

11.5.1 Experiment 1: Autocurricula and Emergent Behavior

In this scenario, we consider a micro-grid of 4 identical production machines (energy loads) executing the same production task. The grid is operated in connected mode. However, the main grid can only provide enough energy for 3 production machines to operate. Therefore, the production machines are in full competition. The exploration reward of the agents is set so that the agent is penalized if the production machine is shut down or suspended. The competition reward is based on the performance of the machines meaning that the agent will receive a positive incentive every time a production task is accomplished. In addition, all agents are penalized if the total energy demand is greater than the energy provided by the main grid. The exploration annealing factor α is set to 0.2 so that the agent will first learn how to execute a production task during the initial 20% of the training time. During the training each agent observed the full environment including the observations and the actions of the other agents. During the execution, the agents only observe the amount of energy available as well as the current operational state of the machine it controls. We train the agents with 2 random seeds for 500 sampling horizons.

Figure 11.7 shows a typical progression of the emergent strategies learned by all agents during the course of the training measured in this case by the respective average cumulative rewards. Clearly, training the agents against each other creates many different strategies, each of which introduces a previously non-existent pressure on other agents to do better in order to move to the next stage. Note that there are no direct incentives for agents to explore, but rather the emergent strategies are solely a result of the auto-curriculum induced by multi-agent competition.

As illustrated in Fig. 11.7, the agents initially learn simultaneously how to control the corresponding production machine so that it can reach the *execute*-state as often as possible. After approximately 30 episodes, all agent have mastered this task and the energy demand of the whole grid begin to surpass the energy provided by the main grid. Consequently, the agents are then frequently penalized leading to a significant reduction of the production performance and the accumulated reward. At this point some agents begin to take advantage of the observations and actions of other agents and intentionally modify their policies accordingly. We can see, for example, some agents wait in the *idle*-state and then switch to the *execute*-state, when other agents are stopped or halted. However, the more the new strategy becomes efficient, the more it becomes predictable and the other agents begin to defend themselves against this strategy. After another 5–20 episodes, all agents have now learned to exploit the strategy of other agents and the energy demand of the production machines again surpass the energy supply and the agents are penalized again more frequently. At this point a new adaptation/exploitation cycle restarts. Because of random exploration factors, the best performing agent during this phase changes from cycle to cycle. After approximately 3–6 cycles (which corresponds to episode 30–160), the training stabilized when at least one agent adopts a very bad policy e.g., only switching between *stopped*- and *idle*- state. The first acceptable coordination strategy emerges from episode 160 to approximately episode 220: *Machine 2*, *Machine 3* and *Machine 4* start to produce more frequently while *Machine 1* remains in the *idle*-state oftenly. It is a valid strategy since the energy required for the production can be supplied by the main grid and the overall productivity is relatively high. The second and final coordination strategy emerge at episode 220: while *Machine 2* and *Machine 4* reached their highest productivity, *Machine 3* is frequently halted.

In this experiment, we found that the emergent auto-curriculum is fairly robust, as similar observations can be made with different initialization strategies. Out of 10 experimental runs, more than 7 runs were able to converge to an equilibrium. However, many design choices can drastically reduce the convergence time and the sample complexity such as the surrogate loss and the entropy loss of the underlying PPO-algorithm. We also experimented with different configurations (for example by including energy storage and generator components) and found that these configurations also lead to emergent behaviors, providing further evidence that multi-agent interaction is a promising path towards self-supervised energy management.

11.5.2 Experiment 2: Fully Cooperative Production

In this scenario, the micro-grid is operated in isolated mode and consists of two production machines (a CNC-machine and a welding machine), an energy generator

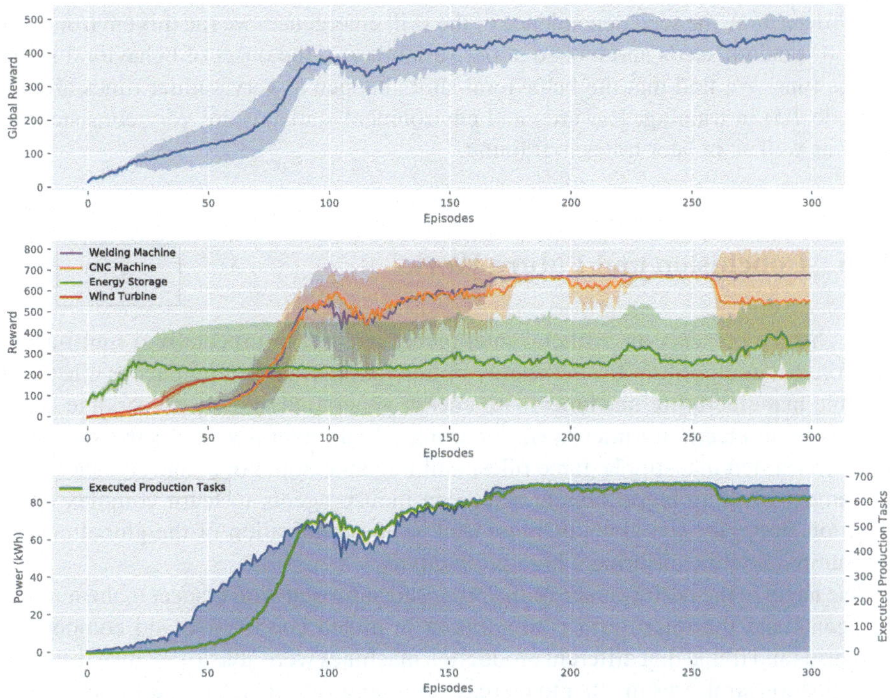

Fig. 11.8 Average rewards on cooperative production. A coordination strategy emerges rapidly after 100 episodes. While the energy generators rapidly learn how to generate energy, the energy storage component adopted to discharge when the production machines in the *execute*-state and thus allowing the production machine to increase the productivity. *Source* Author

(a wind turbine) and an energy storage system. The energy generated by the wind turbine is lower than the energy needed by the production machines. Therefore, the energy storage system has to be turned on in order to execute a production task. All the components should therefore cooperate. The exploration rewards is similar to the previous experiment and the team-based cooperation reward of the agents is based on the performance of the production machines. We train the agents for 300 episodes. The agents also used the same policy network architecture as in experiment 1 (see Sect. 11.5.1).

As we can observe in Fig. 11.8, the components rapidly learn the required basic skills and a coordination strategy emerge around the episode 100. While the energy generator is constantly switched on, the energy storage component learns to charge when the production machines are standing still and to discharge when the production machines perform a production task and thus supports production. Once the energy generator and the energy storage component have adopted a strategy, the production machines rapidly learn to increase their productivity which reach its highest level around episode 250.

To determine the seed dependence of the skill emergence, we ran this environment with 10 random seeds and looked at the variability of the onset of behavioral shifts across runs. We find that the behavioral shifts happen at very similar times around episode 100 in training. Rewards and environment statistics are very consistent in early as well as in later times in training.

11.6 Conclusion and Future Work

This chapter presents an artificial intelligence algorithm especially a multi-agent reinforcement algorithm for the energy optimization of an heterogeneous cluster of flexible manufacturing machines with energy generation and energy storage capabilities in an electricity micro grid featuring a high volatility of electricity prices. It demonstrated that simple game rules, multi-agent competition, and standard reinforcement learning algorithms at scale can induce agents to learn complex coordination strategies and skills that can lead to an optimization of the global energy consumption while ensuring a high productivity.

We analyze the performance of the proposed approach with respect to the number of agents and the micro-grid configuration in mixed cooperative and competitive scenarios and found that different production machines were able to discover different coordination strategies in other to increase the energy efficiency of the whole factory floor. These empirical results are promising and we intend to extend the solution to highly complicated and dynamic environments.

The presented results should be viewed as a proof of concept showing that multi-agent autocurricula can lead to physically grounded coordination strategies. We acknowledge that the strategy space in the proposed environments are inherently bounded and do not reflect the reality of the factory floor since many factors such as product quality, the production time and the unexpected human interventions in the production process have to be considered. However, because the algorithm was built on the functional abstraction of cyber-physical production systems, it is therefore applicable to real production systems if an invocation interface is connected to the agents. Our solution is therefore physically grounded and very extensible. In order to support further research, we have open-sourced our environment code [52].

Acknowledgements The authors would like to thank Joern Pechke, Hans-Henning Klos, Shirin Baer, and others at Siemens AG for interesting discussions related to this paper, as well as Matthias Brossog, Jonathan Fuchs and Dominik Kisskalt for the insightful comments on this manuscript. Finally, we'd like to thank Siemens AG for fostering an engaging and productive research environment.

References

1. Sutton, R., Bach, F., Barto, A.: Reinforcement Learning. A Bradford Book, Massachusetts (2018)
2. Mnih, V., et al.: Playing atari with deep reinforcement learning. CoRR, vol. 13125602 (2013)
3. Aubret, A., Matignon, L., Hassas, S.: A survey on intrinsic motivation in reinforcement learning (2019). arXiv:1908.06976
4. Hausknecht, M., Stone, P.: Deep recurrent q-learning for partially observable MDPs. In: AAAI Fall Symposium on Sequential Decision Making for Intelligent Agents (AAAI-SDMIA15), Arlington, Virginia, USA (2015)
5. Akkaya, I., et al.: Solving Rubik's cube with a robot hand (2019). arXiv:1910.07113
6. Schulman, J., et al.: Proximal policy optimization algorithms (2017). CoRR, arXiv:1707.06347
7. Lowe, R.: Multi-agent actor-critic for mixed cooperative-competitive environments (2017). CoRR, arXiv:1706.02275
8. Bansal, T.: Emergent complexity via multi-agent competition. arXiv (2017)
9. Bai, Y., Jin, C.: Provable self-play algorithms for competitive reinforcement learning (2020). arXiv:2002.04017
10. Tesauro, G.: Practical issues in temporal difference learning. Mach. Learn. **8**(3–4), 257–277 (1992). https://doi.org/10.1007/bf00992697
11. Gholian, A., Mohsenian-Rad, H., Hua, Y.: Optimal industrial load control in smart grid. IEEE Trans Smart Grid, **7**(5), 2305–2316 (2016). https://doi.org/10.1109/tsg.2015.2468577
12. Bakakeu, J., et al.: An artificial intelligence approach for online optimization of flexible manufacturing systems. Appl. Mech. Mat. **882**:96–108 (2018). https://doi.org/10.4028/www.scientific.net/amm.882.96
13. Bakakeu, J.: Multi-objective design space exploration for the integration of advanced analytics in cyber-physical production systems. In: 2018 IEEE International Conference on Industrial Engineering and Engineering Management (IEEM), pp. 1866–1873 (2018)
14. Chen, R., Sun, H., Guo, Q., Jin, H., Wu, H., Zhang, B.: Profit-seeking energy-intensive enterprises participating in power system scheduling: model and mechanism. Appl. Energy 158:263–274 (2015). https://doi.org/10.1016/j.apenergy.2015.08.018
15. Wang Z.: An integrated optimization model for generation and batch production load scheduling in energy intensive enterprise. In: 2012 IEEE power and energy society general meeting, pp. 1–8 (2012). https://doi.org/10.1109/PESGM.2012.6345296
16. Middelberg, A., Zhang, J., Xia, X.: An optimal control model for load shifting—with application in the energy management of a colliery. Appl. Energy, 86(7–8), 1266–1273 (2009). https://doi.org/10.1016/j.apenergy.2008.09.011
17. Sun, Z., Li, L.: Opportunity estimation for real-time energy control of sustainable manufacturing systems. IEEE Trans. Autom. Sci. Eng. **10**(1), pp. 38–44 (2013). https://doi.org/10.1109/tase.2012.2216876
18. Duflou, J., et al.: Towards energy and resource efficient manufacturing: a processes and systems approach. CIRP Ann. **61**(2), 587–609 (2012). https://doi.org/10.1016/j.cirp.2012.05.002
19. Hesselbach, J., Herrmann, C.: Glocalized solutions for sustainability in manufacturing. Springer, Berlin, Heidelberg, pp. 293–298 (2011)
20. Bruzzone, A., Anghinolfi, D., Paolucci, M., Tonelli, F.: Energy-aware scheduling for improving manufacturing process sustainability: a mathematical model for flexible flow shops. CIRP Ann. **61**(1), 459–462 (2012). https://doi.org/10.1016/j.cirp.2012.03.084
21. Fang, K., Uhan, N., Zhao, F., Sutherland, J.: A new approach to scheduling in manufacturing for power consumption and carbon footprint reduction. J. Manuf. Syst. **30**(4), 234–240 (2011). https://doi.org/10.1016/j.jmsy.2011.08.004
22. Liu, C., Xu, X., Hu, D.: Multiobjective reinforcement learning: a comprehensive overview. IEEE Trans. Syst. Man Cybern.: Syst. **45**(3), 385–398 (2015). https://doi.org/10.1109/tsmc.2014.2358639

23. Anderson, C., Hittle, D., Katz, A., Kretchmar, R.: Synthesis of reinforcement learning, neural networks and PI control applied to a simulated heating coil. Artif. Intell. Eng. **11**(4), 421–429 (1997). https://doi.org/10.1016/s0954-1810(97)00004-6
24. Li, L., Sun, Z.: Dynamic energy control for energy efficiency improvement of sustainable manufacturing systems using markov decision process. IEEE Trans. Syst. Man Cybern.: Syst. **43**(5), 1195–1205 (2013). https://doi.org/10.1109/tsmc.2013.2256856
25. Ruelens, F., et al.: Demand response of a heterogeneous cluster of electric water heaters using batch reinforcement learning. In: 2014 Power Systems Computation Conference, pp. 1–7 (2014). https://doi.org/10.1109/PSCC.2014.7038106
26. Ruelens, F., Claessens, B., Quaiyum, S., De Schutter, B., Babuska, R., Belmans, R.: Reinforcement learning applied to an electric water heater: from theory to practice. IEEE Trans. Smart Grid **9**(4), 3792–3800 (2018). https://doi.org/10.1109/tsg.2016.2640184
27. De Somer, O.: Using reinforcement learning for demand response of domestic hot water buffers: a real-life demonstration. In: 2017 IEEE PES Innovative Smart Grid Technologies Conference Europe (ISGT-Europe), pp. 1–7 (2017)
28. Kazmi, H., Mehmood, F., Lodeweyckx, S., Driesen, J.: Gigawatt-hour scale savings on a budget of zero: deep reinforcement learning based optimal control of hot water systems. Energy **144**, 159–168 (2018). https://doi.org/10.1016/j.energy.2017.12.019
29. Jiang, B., Fei, Y.: Smart home in smart microgrid: a cost-effective energy ecosystem with intelligent hierarchical agents. IEEE Trans. Smart Grid **6**(1), 3–13 (2015). https://doi.org/10.1109/tsg.2014.2347043
30. Anvari-Moghaddam, A., Rahimi-Kian, A., Mirian, M., Guerrero, J.: A multi-agent based energy management solution for integrated buildings and microgrid system. Appl. Energy **203**, 41–56 (2017). https://doi.org/10.1016/j.apenergy.2017.06.007
31. Prasad, A., Dusparic, I.: Multi-agent deep reinforcement learning for zero energy communities. arXiv (2018)
32. Lu, R., Hong, S., Zhang, X.: A dynamic pricing demand response algorithm for smart grid: reinforcement learning approach. Appl. Energy **220**, 220–230 (2018). https://doi.org/10.1016/j.apenergy.2018.03.072
33. Littman, M.: Markov games as a framework for multi-agent reinforcement learning, pp. 157–163 (1994)
34. Hu, J., Wellman, M.: Multiagent reinforcement learning: theoretical framework and an algorithm. In: ICML (1998)
35. Busoniu, L., Babuska, R., De Schutter, B.: A comprehensive survey of multiagent reinforcement learning. IEEE Trans. Syst. Man Cybern. Part C (Appl. Rev.) **38**(2), 156–172 (2008). https://doi.org/10.1109/tsmcc.2007.913919
36. Tan, M.: Multi-agent reinforcement learning: independent versus cooperative agents. In: Proceedings of the Tenth International Conference on Machine Learning, pp. 330–337 (1993)
37. Matignon, L., Laurent, G., Le Fort-Piat, N.: Independent reinforcement learners in cooperative Markov games: a survey regarding coordination problems. Knowl. Eng. Rev. **27**(1), 1–31 (2012). https://doi.org/10.1017/s0269888912000057
38. Tampuu, A., et al.: Multiagent cooperation and competition with deep reinforcement learning. PLoS One **12**(4):e0172395 (2017). https://doi.org/10.1371/journal.pone.0172395
39. Gupta, J.: Cooperative multi-agent control using deep reinforcement learning. In: Autonomous Agents and Multiagent Systems, pp. 66–83 (2020)
40. Foerster, J.: Stabilising experience replay for deep multi-agent reinforcement learning (2017). arXiv, Accessed 21 July 2020
41. Lowe, R., et al.: Multi-agent actor-critic for mixed cooperative-competitive environments (2017). arXiv:1706.02275
42. Foerster, J., et al.: Counterfactual multi-agent policy gradients. In: AAAI (2017)
43. Sukhbaatar, S., et al.: Learning multiagent communication with backpropagation. Adv. Neural Inf. Process. Syst. **29**, 2244–2252 (2016)
44. Mordatch, I., Abbeel, P.: Emergence of grounded compositional language in multi-agent populations (2017). arXiv:1703.04908

45. Foerster, J., et al.: Learning to communicate with deep multi-agent reinforcement learning (2016). arXiv:1605.06676
46. Rashid, T., et al.: Monotonic value function factorisation for deep multi-agent reinforcement learning (2018). arXiv:8031.1485
47. Sunehag, P., et al.: Value-decomposition networks for cooperative multi-agent learning based on team reward. In: Proceedings of the 17th international conference on autonomous agents and multiagent systems, pp. 2085–2087. Stockholm, Sweden (2018)
48. He, H., et al.: Opponent modeling in deep reinforcement learning. In: Proceedings of the 33rd international conference on international conference on machine learning, vol. 48, pp. 1804–1813, New York, NY, USA (2016)
49. Choi, J., Lee, B., Zhang, B.: Multi-focus attention network for efficient deep reinforcement learning (2017). arXiv:1712.04603
50. Jiang, J., Lu, Z.: Learning attentional communication for multi-agent cooperation. In: Proceedings of the 32nd International Conference on Neural Information Processing Systems, pp. 7265–7275. Canada, Montreal (2018)
51. Brockman, G., et al.: OpenAI Gym (2016). arXiv:1606.01540
52. Bakakeu, J.: Multi-Agent-RL-Energy-Management, GitHub Repository (2019). https://github.com/jupiterbak/Multi-Agent-RL-Energy-Management

44. [illegible] Learning to communicate with deep multi-agent reinforcement learning. CoRR, abs/1605.06676 [illegible]
45. [illegible] 2018 [illegible]
46. [illegible]
47. [illegible]
[illegible]
[illegible]
[illegible]

Chapter 12
Affect of Artificial Intelligence Technologies and Digitalisation on Jurisprudence and Education

Tatiana V. Emelianova

Abstract The chapter investigates the issues of artificial intelligence technologies /AIT/ and innovations in digital technologies affecting jurisprudence and education. Technologies associated with artificial intelligence like advanced computing, artificial neural networks, big data, machine learning, unmanned systems and robotics give new impetus to the development of universities and educational institutions around the world. The goal of this research is to demonstrate that Masters of Arts studying at RUDN University are trained for the English-language law activities affected by digitalisation: cyber security and culture, preventions of sexual offences against children facilitated through the use of information and communication technologies (criminal law), smart contracts and arbitration (civil law), e-filing—filing documents electronically with the courts, etc. The survey of students having legal practice simultaneously with the studies proved that they are leveraging AIT and digitalisation as a key strategic asset. The problems of AIT implementation in contemporary lawyers' activities like ousting legal professionals from law services market by AIT are analysed. The chapter presents some recommendations on making legal practices more efficient by speeding up and facilitating the work in legal activities based on digital technologies. This research results can be applied by scientists and practitioners involved in AIT, educational and legal processes.

12.1 Introduction

The environment we operate in is constantly evolving; digital technologies and artificial intelligence influence the world faster than we could predict. Lawyers, educators, administrators—we all have no choice but to adapt and respond to these changes by undergoing digital transformation. Going digital is not only a challenge, it is an unprecedented opportunity. Innovative solutions can help various processes function more efficiently.

T. V. Emelianova (✉)
RUDN University, 6 Miklukho-Maklaya Str, Moscow, Russian Federation
e-mail: emtv@mail.ru

A. Dingli et al. (eds.), *Artificial Intelligence in Industry 4.0*,
Studies in Computational Intelligence 928,
https://doi.org/10.1007/978-3-030-61045-6_12

Every day the Internet is increasing more and more in its scale; a huge number of law websites and online legal services are being created to provide various types of information. They open up new opportunities in communication and the transfer of knowledge over long distances. Our century is the era of new progressive technologies, which successfully intervene in all sectors of human activities including the legal services market.

All digital technologies and Artificial Intelligence Technologies /here and after AIT/, that may reduce time of work and increase production efficiency in legal service sector, are called "Legal Technologies" or just "LegalTech". This chapter provides an overview of the phenomenon of AIT, digitalisation, blockchain, LegalTech, cybercrime and cyber culture as global problems. Their features and possible development tendencies are revealed. The goal of this research is to demonstrate that masters of Arts studying at the Institute of Law, RUDN University are entirely trained in the field of the English-language professional interaction and law activities affected by digitalisation: blockchain, cyber security and cyber culture, preventions of sexual offences against children facilitated through the use of information and communication technologies (criminal law), smart contracts and smart arbitration (civil law), e-filing—filing documents electronically with the courts, etc. Digitalisation offers new ways of interacting with taxpayers by allowing administrations to harness data to offer customers new proactive services; moreover, digitalisation makes administrations more agile in their compliance activity, boosting their performance. The technology of blockchain has a great potential.

The problems of digital technologies and their application in the activities of modern lawyers, such as ousting lawyers from the legal services market by artificial intelligence technologies, are analysed. The methodological basis of this study is the research and fundamental works of Russian and foreign scholars and practitioners in the field of jurisprudence, legal linguistics, theory and methodology of teaching foreign languages (level of professional education), in the field of information and digital technologies.

The material presented in this work is the result of theoretical and practical activities of the author of this article. By the method of comparison and grouping, in 2017, 2018, and 2019, the study analysed a series of surveys of fifty-eight (58) masters of Arts aged 21–26 years to identify the importance of learning the models of the English-language professional interaction and law activities affected by AIT and digitalisation. The data revealed that 85% of respondents, having legal practice simultaneously with the studies at the University, were leveraging artificial intelligence technologies and digitalisation as a key strategic asset. Digitalisation, AIT and LegalTech are not a threat to legal services' market. Although these innovations are changing legal service market, it is not possible that computer technologies or AIT may totally replace a lawyer in this market in future.

12.2 Related Works

The processes of Industry 4.0, rapid growth and development of artificial intelligence technologies and Internet resources may have both negative and positive aspects; everything depends on the ultimate goal: virtual personal assistants like Siri, Alice, Cortana, video games in which AI is used to create characters, cars with artificial intelligence, smart homes and more, Skype, ZOOM and Microsoft Teams are great tools for running online lessons for the students of masters' programmes who choose distance education.

Industry 4.0, in particular, artificial intelligence technologies have an effect on the contemporary global processes and form new approaches to educational and legal technologies. The system of vocational education is posed to a certain challenge. Modern civilisation actively develops, expands, and implements numerous inventions of artificial intelligence and digital technologies. It involves the infusion of data and smart automation within not only the manufacturing process but in all spheres of life and this process develops in high gear.

The concepts and doctrinal approaches of researchers in the sphere of jurisprudence, linguistics, theory and methodology of teaching foreign languages and vocational training are the methodological bases of this work. The publications of international scientists in the field of Industry 4.0, artificial intelligence, law, English for special purposes, intercultural interaction, and forensics have formed the theoretical basis of this study.

Various international scientists, practitioners, and academicians in their research and surveys focus on the disputable issues of indispensability and significance of Industry 4.0, artificial intelligence and digital technologies, law, education, and their interrelation in some cases [1–3, 5, 9, 11–14, 19, 21, 23, 24, 26–30].

The feasibility and thus inevitability of the English-language advanced curriculum at non-linguistic universities in the context of subject-specific (professional) and general cultural competences was emphasised [7, 8].

The main scientific methods like analysis, synthesis, systematisation of research material, comparative legal and formal legal methods helped to investigate the norms of the current Russian and International legislation that regulate the educational process and the actual relations that arise within Industry 4.0.

Industry 4.0 and AIT regulation is not a specialised sub-discipline, it affects the entire educational and legal systems and therefore concerns all educators and lawyers. Legal regulation is considered important both inside and outside of legal science where such a phenomenon as legal logic becomes apparent. It is the type of logic that operates in the context of social conditions, government policies, and groups of people. This issue has become even more important with the introduction of artificial intelligence into the legal and educational systems.

Excessive accessibility, weak control over data and the possibility of circumventing a huge number of prohibitions make the Internet a very convenient tool for various criminal groups, thus giving rise to these new global problems of cybercrimes. Many countries and international organisations, in particular, the United Nations Organisation, pay much attention, issuing various decrees and legal documents calling for the development of measures to combat the global threat.

In the history of cybercrime, several stages of development can be singled out: the development and dissemination of a global Internet network; large hacker groups emerged in the late 90-ies; during the 2000s, the activities of cybercriminals acquired a transnational nature with a specialisation in cyber terrorism and cyber war, etcetera.

We are on the so called fourth stage of the cybercrime's development up to date. In general, cybercriminals direct their activities to financial and political infrastructure. Consulting Legal Dictionary [25], cybercrime is currently defined as any criminal offense facilitated or incorporated by the use of electronic communications or information systems, including any electronic device, computer or Internet.

Attacks from computer systems and networks are spontaneous and unpredictable. It is almost impossible to predict where and in what form the threat will come from, which reduces the chances of its elimination. Particular attention was given to combatting the use of information and telecommunications systems for criminal purposes at the VII Conference of Attorneys General of Europe held in 2006 [18], the need to develop a global strategy against terrorism was emphasised there.

According to the statistics presented by the General Prosecutor's Office of the Russian Federation in 2018 [10], the number of crimes committed through the use of modern information and communication technologies increased from 65,949 to 90,587 in 2017. Their share in all criminal acts registered in Russia is 4.4%; this is almost every twentieth crime.

Together with this, crimes in the networks take special positions: the facts of high rates of cybercrimes' growth and misuse of AIT and digitalisation around the world are noted by influential authoritative sources as one of the most important problems of the modern world.

These problems are regulated by various legal documents like Articles 272, 273, 274 Chap. 28 of the Criminal Code of the Russian Federation dated 13.06.1996 No. 63-FZ the latest edition of which came into force in 2018 [6] or some materials of the Lanzarote Convention—Council of Europe Convention published in 2014 [4] where special attention was paid to the protection of children against sexual exploitation and sexual abuse and the prevention of crimes of a sexual nature against children committed with the use of information, communication and digital technologies (Internet).

Like humans, artificial intelligence can also make mistakes; in the case of people, it is clear who has to be responsible for their mistakes or misuse of AIT. When we talk about law, mistakes can lead to appalling consequences. Accordingly, the question arises, whom do we have to judge to be responsible for the errors of the "robot"? This issue should be resolved by legislation by introducing new legal norms that have not existed before, because there was not so well-developed artificial intelligence.

The eradication of these problems requires effective cooperation of all countries. In this regard, the prevention of crimes in the digital environment and the elimination of their consequences are truly global. Thus, masters of Arts studying at Law

Institutes or Universities must be well educated in the context of the English-language professional interaction and law activities affected by digitalisation within the processes of Industry 4.0.

12.3 Proposed Work

This chapter presents some recommendations to improve the efficiency of professional English-language communication in legal practice by accelerating and facilitating the work in various legal activities with the use of AIT and digitalisation. These recommendations are based on the materials which are the results of theoretical and practical activities of the author of this work. The survey of 58 masters of Arts having legal practice simultaneously with the studies at the Institute of Law, RUDN University proved that they are leveraging artificial intelligence technology and digitalisation as a key strategic asset. Thus, within the studying years of 2017, 2018 and 2019 all masters, including those studying the English language within the master's programme "Legal Regulation of Blockchain Technology", were asked several questions, for example:

- Is the English Language the so-called stimulus to progress regarding Industry 4.0?
- Do you think that the English Language is one of the main elements of Industry 4.0?
- Is it important to know the English Language for the lawyers who deal with products of artificial intelligence?
- Do you want to know the basic set of words in the English language for the regulation of digital technologies and AI?
- Is it important to understand the main legislative acts in the field of regulation of digital technologies and AI and compare them with the legislative acts of foreign countries?
- Do you know any key concepts on the regulation of digital technologies and AI?
- How does knowledge of the English language help in studying the regulation of digital technologies and AI?
- Do you consider that it is necessary to widely introduce digital technologies and AI into the sphere of public regulation?
- Is the States' strong legislative regulation of digital technologies necessary?
- In anticipation of the upcoming digital revolution in public administration, how should business participants and government representatives act to successfully introduce new technologies?
- In what period of time will financial control be carried out mainly by AI and digital technologies?
- Why is it better for students to be encouraged to explore digital opportunities now?
- Do you think that the technological revolution, in particular, the introduction of AI into existing relations will affect the decline of jobs?

The analyses of the answers of masters helped to reveal the strong necessity to study the issues of AI and digital technologies in the English language in interdisciplinary context, and to train them in the field of law activities affected by digitalisation.

Throughout the process of teaching actual professional English-language interaction to the masters of Arts, students are offered numerous authentic texts of relevant topics for reading and translating from English into Russian and vice versa, cases from modern practice are considered and solved, the method of projects is used, debates on these problematic issues are held.

The ongoing discussions considering the legal status of the robots still arise interest among masters of Arts. The author of this research provided them with an argumentation activity in classroom during which they discussed the prerequisite to determine legal status of electronic machines and their role in society since they work closely with people. This concerns those robots that have artificial intelligence and are able to move independently, in particular, humanoid robots and unmanned vehicles. The students also discussed the pros and cons of the necessity to take into account the contribution of robots to the development of new economic paradigm.

The summary of this activity culminated in a strong believe, in favour of creating a legislative council to ensure the development of the digital environment, which will include lawyers as well as experts from the IT industry. The students correspondingly stated that each "smart" machine must be identified; its data must be entered into a specially created registry.

Throughout the discussions run during classroom activities, there were also dubious opinions on the issues of AIT and digitalisation, for example, in the near future, carriers of AI—the so-called "electronic persons", may acquire certain rights. If so, they will be able to be subjects of law, and, quite possibly, will be able to participate in public relations on an equal basis with people. Many aspects of the development of AI will be hidden from plenty of people until the products of AI are investigated and learned empirically as to accurately predict the consequences of the development of AI is impossible. This may be the acceleration of the progress of mankind, which will solve many global problems, but it can also bring others right up to the collapse of the human age.

Throughout the process of teaching the English language for masters' implementation in factual professional legal activities, they have to be informed that there are many spheres of jurisprudence in which free creativity plays only a subordinate role; the stereotypical rules of behaviour mainly concern registration of most legal entities and property rights, imposition of fines and arrests, receiving various kinds of statements, filing tax returns and reports, making payments, conclusion of simple contracts. In these areas, above all, digitalisation will still occur and keep rather stable positions.

When training masters of Arts for actual professionally oriented English-language interaction and implementation in legal activities connected with digitalisation processes, it is emphasised that the most common types of cyber-crimes are unauthorised access to computer information. Thus, masters of Arts are recommended to read, translate from Russian into English, and after that present throughout classroom activities the main provisions of articles 272, 273 and 274, Chap. 28 of the Criminal Code of the Russian Federation [6], which regulate these problems, in particular, illegal access to computer information; creation, use and distribution of malicious software; violation of the rules for means of storage operation, processing and transfer of computer information and information and telecommunication networks. The search for similar regulating and legal English-language documents in the United Kingdom and the United States, followed by a comparison of the data and discussion during the practical lessons, expands the vocabulary and knowledge of students in this field of law.

Some materials of the Lanzarote Convention—Council of Europe Convention [4] where special attention is paid to the protection of children against sexual exploitation and sexual abuse and the prevention of crimes of a sexual nature against children committed with the use of information, communication and digital technologies (Internet) are being translated. The subsequent thoughtful discussion of this problem in the English language during classroom activities develops masters of Arts' critical thinking and creates prerequisites for gamification—that is the so-called gaming lawmaking throughout practical classes with the development of amendments to laws, etc.

Another useful task is to ask them to participate in research projects, for example, search for statistics on reported crimes and to prepare for debates. Thus, for example, if in 2017 there were registered 1,883 crimes of this type (+7.7%), after that, in the first half of 2018 there were registered 1,233 crimes (+3.4%) as it was presented by the General Prosecutor's Office of the Russian Federation in 2018 [10].

This information is then discussed in the English language during classroom activities. Masters of Arts are invited not only to participate in the debate, but also to make suggestions and recommendations to address the problem from their perspective, based on their knowledge and experience in the legal field; at the end of the debate, students are recommended to write a so-called communiqué in English.

All of the above involves the active use of various dictionaries like Legal Dictionary [25], Cambridge Dictionary online [20], Longman Dictionary of Contemporary English [22], English-Russian Comprehensive Law Dictionary [16], Explanatory dictionary of the Russian language [17], and many others. They allow masters to adequately apply professionally directed English vocabulary in speech and writing.

Another important practical activity with the students is the consideration and discussion of the professional activities of criminalists through the prism of new technologies that gives impetus to the disclosure and efficiency of the investigation staff. The effectiveness of the investigation of crimes essentially depends on the array of forensic information that the investigator has and the people involved in interaction. In recent years, modern information technologies that rely on sufficiently powerful computers equipped with appropriate software, as well as computer systems and networks, have been increasingly used to collect and process this type of information.

Modern tools and methods for the implementation of information processes, especially the Internet, create a kind of special virtual reality, presenting a specific picture of reality through the transmission of messages, images, texts, etc. Information and analytical work in the investigation of crimes includes the collection,

storage,
systematisation and analysis of evidence and guidance information in order to make optimal tactical and procedural decisions. The use of modern information technologies not only rationalises information processes occurring in criminal proceedings, but also contributes to making effective decision by investigators, experts, operatives, judges supported by all types of digital systems.

It is impossible to increase the effectiveness of law enforcement agencies' work in the detection and investigation of crimes without integrating new information technologies into the criminalistics. In no way inferior to the capabilities of their predecessors—large-scale computers, personal computer /here and after PC/, especially laptops and tablets have, in comparison with them, such undoubted advantages as relatively low cost, compactness, high reliability, efficiency, etc.

This allows lawyers to implement them literally at every workplace both autonomously and included in local information systems and networks, or as terminals. When investigating a particular criminal case, information about the mechanism and method of committing a crime, the subject of criminal infringement, the victim, etc. is entered online in PCs. After processing, the monitor exposes recommendations that can be used in planning the investigation, it allows grouping data by episode and participant, and also carries out a specific investigative action, search and comparison of episodes, surnames, nicknames, dates, etc.

Computer "intelligence" is in this case a generalised best practice of investigative, expert activity, and the issued recommendations make it possible to optimise decisions made. Thus, masters of Arts are asked to prepare and decide their own cases (based on real ones) implementing the above information and various digital technologies; they present the results in classroom activity to the group-mates, which is followed by the analysis and discussion with the students, the role of the teacher is to correct all mistakes (grammar, phonetics, lexis and terminology, stylistics, content and data, etcetera). Taking into account the above, it can be concluded that new digital technologies that are being improved day by day give a more effective assessment of what is happening and the possibility of a detailed investigation of a particular issue thus bringing to the necessary results.

12.4 Discussion

It is impossible to fully automate either lawmaking or law enforcement. Artificial intelligence can be used in these areas, but it should be of a secondary role in helping a person to discover the contradictions of the norms of law; observe their duplication (redundancy) or illogicality. We still have decisions about what these norms are as well as how, in the end, they should be applied by a person. This does not mean that the idea of digitalisation of law does not make sense at all.

The main thing is not to absolutise it, not to become its indiscriminate propagandists. There are some affordable and available LegalTech products on the legal market, such as document-builders, document-constructors, document analysis and research programmes. These products have a massive potential for the reduction of lawyers' working hours.

Talking about blockchain technology, its potential is real-time, immutable, decentralised, trusted and transparent transactions—exactly the sort of characteristics that are becoming ever more important in the global digital tax system. The international push for tax transparency and inclusive action against tax avoidance has created uncertainty around international tax positions, and resulted in major compliance burdens on bigger businesses. Blockchain has the potential to simplify and accelerate transparency compliance, tax audits, and decision making. It may also be the key to facilitating collaboration among business, tax authorities, and professional advisors.

Blockchain has already revealed a lot of benefits and while the big buzz is around the financial domain and banking sector, from a long time point of view, it will also reveal its potential in the taxation world. Given all these factors, it is reasonable to state that it was a very proper step for law students to start studying and researching the problems of digitalisation of taxation. In order to be aware of the terminology used throughout discussed issues with the students and, finally, speak the lingo, it was really important to surf the internet and to pick out internationally recognised key language, thus compiling a glossary, for example:

- ASIC is an integrated circuit designed specifically to perform a special task, in this case for mining, which can be done much more efficiently than a Graphical Processing Unit (GPU) or a Field Programmable Gate Arrays (FPGA). The main disadvantage of this equipment is its inapplicability to other tasks, which seriously increases the risk of losing money.
- Blockchain is a continuous sequential chain of blocks (linked list) containing information built according to certain rules.
- Block is a data structure that contains all the necessary metadata about the block itself (block header) and contains transactions. The first block in the blockchain is called the genesis block.
- Bitcoin is the first decentralised cryptocurrency based on open source code and functioning on the p2p protocol.
- Confirmation is the act of successfully adding a transaction to the blockchain after verification; as a rule, more evidence means more protection against double attack (persistence).
- Consensus is the decision to include a specific block in the blockchain, the achievement of which is possible only after verification. There are various consensus building mechanisms.
- Cryptocurrency is a digital asset.
- Digital Signature is a code generated from a public key that allows you to evaluate whether the addressee really wants to conduct a transaction.
- Distributed book is a register that is supported on many nodes in a decentralised network. Records are stored in chronological order. This book can be of two types: allowed and not allowed, depending on who has access to view the book.
- Double expense is the act of using the same digital asset (cryptocurrency) twice. This is a common type of attack on blockchains. This type of attack becomes more

complex with the increase in the number of participants who add a new block to the chain.

- Mining is the term for the process of finding new blocks were not accidental: involuntarily associations with the gold mining process arise. In our case, everything looks as follows: many computers compete for the fastest finding a suitable hash function that allows you to save transactions in the next block; for finding a solution, the miner is entitled to a reward of 12.5 BTC, which makes this process cost-effective.
- Orphan Block is a block whose ancestor was not processed by a node. It cannot be included in the network, as it is contained in a shorter blockchain branch.
- Ring signatures is a method of confirming the authenticity of transactions, which allows not to disclose the sender.
- Smart contract is a computer algorithm designed to generate, control and provide information about ownership of something.
- Transaction is an exchange of assets between two parties or organisations.
- Transaction processing fee is a contribution made by the sender for faster processing of a transaction on the blockchain. The higher the size of the commission, the faster it will be recorded in the registry.
- The operator of the investment platform is a business company that provides access to the investment platform to its participants, as well as carries out activities to organise investment attraction. This person establishes and discloses to the participants the rules of the investment platform, the content requirements of which are determined by law. In accordance with these rules, the operator enters into agreements on the provision of services to attract investment and on the provision of services to facilitate investment with persons attracting investments and investors, respectively.
- Wallet is a file that generates and contains your private keys.

Reading, translating and discussing the experiment "Artificial Intelligence versus Lawyers" [15] was really useful throughout classroom activities with masters of Arts. The experiment showed that LegalTech does simple legal work faster than a lawyer; for example, there was an experiment in the United States of America in which several highly-qualified lawyers and one AI analysed five different Non-Disclosure Acts against time. Total average time spent by lawyers was 92 min but it took only 26 s for AI to do the same. At least this fact proves the statement about LegalTech higher productive efficiency compared to human beings. However, all LegalTech experts agreed that LegalTech is unable to deal with extraordinary tasks as solving complicated and difficult legal issues or cases.

Time savings as the mostly talked about benefit of applying artificial intelligence in law which translates, in turn, to monetary savings were also stated by Kevin D. Ashley in 2017 in his book [2] that was also recommended to masters of Arts for reading, translating and commenting. He also accentuated, that computers are simply exponentially faster than human beings at processing certain types of information, especially when it comes to specific, narrow tasks like searching through and identifying relevant data, conducting legal research and analysing documents for errors, gaps in information and inconsistent language faster than a human being.

The field of education, along with a number of other industries (such as health and telecommunications) is subject to significant changes due to the increasingly active dissemination of digital technologies. Modern digital technologies provide new tools for the development of universities and other educational institutions around the world. Digitalisation provides opportunities for the exchange of accumulated experience and knowledge, which allows people to learn more and make more informed decisions in their daily lives.

Among the interesting digital innovations, it should be noted the rapid adaptation of online learning, which is expressed in the form of the development of mixed forms of teaching (blended learning) and in the active development of online courses MOOC (Massive online open course). The dynamics of the development of online learning is demonstrated, in particular, by the growth of available online courses, the number of which has doubled annually in recent years—more than 4200 courses are offered now by more than 500 universities. Additional areas of application of digital technologies in education are the development of digital libraries and digital campuses of universities, which have already been implemented by many universities in Asia, America, Europe, and Russia.

Thanks to digitalisation, today everyone can access information that was previously only available to experts and scientists. The world of education and science has become global, it is now almost impossible to find a student, a teacher or a scientist who would not have visited foreign universities as part of academic mobility programmes. We live at an interesting time, when the concentration of new digital technologies is greater than ever. In the future, there will be more and more technologies for high-quality and fast information retrieval, and for us, both the educators and students, this is undoubtedly an excellent opportunity to realise our capabilities and potential in general.

In the new digital era, this type of data and knowledge will be the most valuable asset; the students revising these problems, in addition to general professional literacy, will also advance in language competence, as most of the studied documents, research materials, and literature belong to foreign authors. These students will be able to stand out qualitatively against the background of their colleagues, for they will become the so-called "Christopher Columbus" of artificial intelligence and digital technologies. If the process cannot be stopped, it is worth leading.

12.5 Conclusions

Today we witness a new era of technological breakthrough that changes the existing paradigms, stimulates the creation of new orders and adoption of new laws. The existing driving force which is called artificial intelligence and digitalisation has already a strong impact on law, education, the economy, politics, medicine and many other fields of human activities. There is no doubt that in future it will only increase its presence in a variety of areas of life.

The study of artificial intelligence and digital technologies in general is very useful for contemporary students as the emergence and application of new tools always require the presence of specialists capable of using them as well as regulating and managing them where the legal aspects of AI & DT execution prove to be extremely actual. With every change and advance, we see these new forces testing the strength of the existing regulations. Moreover, the existing regulations are not always ready for new technologies, which is why it is especially valuable for the students studying law in the English language, to advance their knowledge of International and Russian legal regulation of artificial intelligence and Industry 4.0, and to master the directions of digital transformation.

The implementation of artificial intelligence technologies and innovations in all digital technologies help find conflicts in laws. These huge layers of identified problems in legal practice at the moment, due to limited human capacities, require large resources to analyse the contradictions and improve the legislative framework. These types of innovations will certainly help lawyers to correctly interpret and resolve cases, and assist clients. To keep in step with the times, it is indispensable to educate masters of Arts to implement these technologies as they should be applied and adapted in all spheres of human activities.

Nevertheless, neither LegalTech nor artificial intelligence technologies may entirely replace a lawyer in legal compliance market, due to technological lack of availability in solving complicated and extraordinary issues or cases, but a high risk of unemployment for the beginners in legal compliance market is a real threat. Legal firms have no need to hire and pay young unskilled lawyers as LegalTech may complete all routine simple work incredibly fast and effective. In fact, we are facing no threat to legal services market as it is today, but we are facing a massive problem with large amount of unemployed young specialists in future.

That is why it is a doubtful statement, that the majority of lawyers must be entirely trained in the field of law activities, including the English-language professional interaction, affected by digitalisation. The studied information throughout classroom English-language activities regarding blockchain, cyber security and cyber culture, preventions of sexual offences against children facilitated through the use of information and communication technologies (criminal law), smart contracts and smart arbitration (civil law), filing documents electronically with the courts, the other problems covered above, and implemented methods of teaching (project work, case-study, deliberations, role-plays, gamification, etc.) enlarged graduate-masters' chances for their competitiveness in the labour market thus for being successfully employed. The issues described in this paper may be useful for scientists, researchers, starting the career and practicing educators of Law Institutes and Universities.

12.6 Future Work

Many multifaceted issues still need to be addressed and solved, obviously, students who are now exploring the introduction of artificial intelligence and digital technologies into different spheres of life, will undoubtedly soon be one step ahead of those indifferent; many options for the use of artificial intelligence and digital technologies, including blockchain, provide the widest opportunities for those students who study at the moment. The author of this research strongly believes that accomplishments will be achieved by those who are in the forefront and are able to make the most of all the benefits of digital transformation.

Artificial intelligence technologies and digitalisation allow not only to analyse the past and the present, but also to determine the future. Based on the predictive analysis of big data, legal educators and lawyers can, for example, build their own case management strategies or decide on a pre-trial dispute. With regard to judicial practice in future, it looks like to be in need of judicial analysts who can quickly navigate the latest expert-analytical systems. The computer "vision" is able to accurately determine the emotions of the defendant, the witnesses, and thus facilitate the work of the jury and prevent pressure attempts, as well as collusion.

References

1. Antonopulos, A.: Mastering Bitcoin: Programming the Open Blockchain, 2nd edn., pp. 1–396. O'Reilly Media, Sebastopol (2017). ISBN 978-1491954386
2. Ashley, K.: Artificial Intelligence and Legal Analytics: New Tools for Law Practice in the Digital Age, pp. 1–426. Cambridge University Press, Cambridge (2017). ISBN 9781316622810
3. Bellantuono, G., Lara, F.T. (eds.).: Law, Development and Innovation, pp. XII, 215. Springer International Publishing, Switzerland (2016). ISBN 978-3-319-13310-2. https://doi.org/10.1007/978-3-319-13311-9
4. Convention, C.E.: Council of Europe Convention: Protection of Children against Sexual Exploitation and Sexual Abuse. Strasbourg, Council of Europe. ISBN 978-92-871-7572-4. https://rm.coe.int/protection-of-children-against-sexual-exploitation-and-sexual-abuse/1680794e97 (2014). Accessed July 2014
5. Corrales Compagnucci, M., Fenwick, M., Forg ó, N. (eds.).: New Technology, Big Data and the Law, pp. XVI, 330. Springer, Singapore (2017). ISBN 978-981-10-5037-4. https://doi.org/10.1007/978-981-10-5038-1
6. CriminalCode, R.F.: The Criminal Code of the Russian Federation dated 13.06.1996 No. 63-FZ in the latest edition, in force from October 21, 2018. http://logos-pravo.ru/ugolovnyy-kodeks-rf-uk-rf (2018). Accessed January 2019
7. Emelyanova, T.V.: Lawyer's communicative competence formation for intercultural professional interaction. In: 4th International Multidisciplinary Scientific Conference on Social Sciences and Arts SGEM 2017, Conference Proceedings, 24–30 August, 2017, Book 3, vol. 4, SGEM, Bolgaria, pp. 339–346 (2017a). ISBN 978-619-7408-21-8/ ISSN 2367-5659, https://doi.org/10.5593/sgemsocial2017/34/S13.044, https://sgemsocial.org/ssgemlib/spip.php?article4277&lang=en (2017). Accessed Jan 2018
8. Emelyanova, T.V.: Peculiarities of teaching legal English in multinational groups. In: Proceedings from 7th International Conference on Intercultural Education "Education, Health

and ICT for a Transcultural World", EDUHEM 2016. Procedia – Social and Behavioral Sciences, vol. 237, pp. 1494–1499. Elsevier Ltd., Almeria, Spain (2017b). https://doi.org/10.1016/j.sbspro.2017.02.235 (2017). Accessed 21 Feb 2017
9. Franco, P.: Understanding Bitcoin: Cryptography, Engineering, and Economics. Wiley Finance Series, Wiley. pp. 1–291 (2015). ISBN 9781119019169
10. Genproc.gov.r: General Prosecutor's Office of the Russian Federation. [online] On crimes committed with the use of modern information and communication technologies. https://genproc.gov.ru/smi/news/news-1431104/ (2018). Accessed 14 Aug 2018
11. Hallevy, G.: Liability for Crimes Involving Artificial Intelligence Systems. Springer International Publishing, pp. VIII, 257 (2015). ISBN 978-3-319-10123-1, https://doi.org/10.1007/978-3-319-10124-8
12. Hilbert, M.: Quantifying the Data Deluge and the Data Drought (April 1, 2015). SSRN: https://ssrn.com/abstract=2984851 or http://dx.doi.org/10.2139/ssrn.2984851 (2015). Accessed Jan 2019
13. Hilbert, M., L 6pez, P.: The World's Technological Capacity to Store, Communicate, and Compute Information. Science, vol. 332, Issue 6025, 01 April, 2011, pp. 60–65. https://doi.org/10.1126/science.1200970. http://science.sciencemag.org/content/332/6025/60 (2011). Accessed Jan 2019
14. Jerry Brito et al.: The Law of Bitcoin, Kindle Edittion. iUniverse. Bloomington, IN, pp. 1–228 (2015). ISBN-13: 978-1491768686
15. LawGeex.: Artificial Intelligence versus Lawyers: the ultimate showdown. https://www.lawgeex.com/AIvsLawyer/ (2018). Accessed Jan 2019
16. Mamulyan, A.S. (ed.), Kashkin, S.Yu.: English-Russian Comprehensive Law Dictionary. Academic Edition. Moscow, Russia: Eksmo (Russian legal education), pp. 1–816 (2008). ISBN 978-5-699-11196-1
17. Ozhegov, S.I., Shvedova, N.Yu.: Explanatory dictionary of the Russian language. 4th edn. Moscow, Russia: ITI Technologies, pp. 1–944 (2006). ISBN 5-902638-10-0
18. Putin, V.V.: Concluding remarks at the Seventh Conference of European Prosecutors-General. Published in section: Transcripts. Publication date: July, 5, 2006. http://en.kremlin.ru/events/president/transcripts/23688 (2006). Accessed Jan 2019
19. Radu, R., Chenou, J.M., Weber, R.H. (eds.).: The Evolution of Global Internet Governance. Principles and Policies in the Making, pp. XLIX, 198. Springer, Berlin. (2014). ISBN 978-3-642-45298-7, https://doi.org/10.1007/978-3-642-45299-4
20. Reference.: In: Cambridge Dictionary [online]. Cambridge University Press. https://dictionary.cambridge.org/ (2019). Accessed 9 Jan 2019
21. Sartor, G., Branting, L. (eds.).: Judicial Applications of Artificial Intelligence, pp. IV, 222. Springer, Netherlands. (1998). ISBN 978-0-7923-5472-7, https://doi.org/10.1007/978-94-015-9010-5
22. Summers, D.: Longman Dictionary of Contemporary English, pp. 1–1950. New ed. Pearson Education Limited, Longman (2003). ISBN 0-582-50664-6 (pbk)
23. Swan, M.: Blockchain: Blueprint for a New Economy, pp. 1–129 . O'Reilly Media, Inc. (2015). ISBN 978-1491920497
24. Tapscott, D., Tapscott, A.: Blockchain Revolution: How the Technology Behind Bitcoin Is Changing Money, Business, and the World Portfolio. Penguin, pp. 1–368 (2016). ISBN 978-1101980132
25. Team, C.: Cybercrime – Definition, Examples, Cases, Processes. [online] Legal Dictionary. https://legaldictionary.net/cybercrime/ (2019). Accessed 9 Jan 2019
26. Thames, L., Schaefer, D. (eds.).: Cybersecurity for Industry 4.0. Analysis for Design and Manufacturing, pp. XIII, 265. Springer International Publishing (2017). ISBN 978-3-319-50659-3, https://doi.org/10.1007/978-3-319-50660-9
27. Walton, D.: Argumentation Methods for Artificial Intelligence in Law, pp. XVIII, 270. Springer, Berlin, (2005). ISBN 978-3-540-25187-3, https://doi.org/10.1007/3-540-27881-8
28. Weber, R.H.: Realizing a New Global Cyberspace Framework. Normative Foundations and Guiding Principles, pp. XXVII, 166. Springer, Berlin (2015). ISBN 978-3-662-44676-8, https://doi.org/10.1007/978-3-662-44677-5

29. Weber, R.H., Staiger D.: Transatlantic Data Protection in Practice, pp. XXXII, 162. Springer, Berlin (2017). ISBN 978-3-662-55429-6, https://doi.org/10.1007/978-3-662-55430-2
30. Wischmeyer, T., Rademacher, T. (eds.).: Regulating Artificial Intelligence, pp. XIV, 388. Springer International Publishing (2019). ISBN 978-3-030-32360-8, https://doi.org/10.1007/978-3-030-32361-5

Part IV
Agents in Industry 4.0

In 30 years, a robot will likely be on the cover of TIME magazine as the best CEO. Machines will do what human beings are incapable of doing. Machines will partner and cooperate with humans.

Jack Ma
Alibaba

Chapter 13
Approach for Model Driven Development of Multi-agent Systems for Ambient Intelligence

Tsvetelina Ivanova and Idilia Batchkova

Abstract Ambient Intelligence systems are smart environments, which are reactive and proactive to people making their actions safer and more efficient. In respect to the relation with users, they should be non-obtrusive, context aware, personalized, adaptive and anticipatory. Reliable way to achieve these main features is the choice of appropriate development approach that will ensure the reusability, portability, scalability and interoperability of ambient intelligence applications. The model driven development is a promising approach for development of different software applications using models at different levels of abstraction and applying model transformation to code generation. In this paper an extension to the traditional model driven development approach is suggested based on model driven architecture and compositional models as core elements for automation of expert activities, enabling the context-awareness of application. The ambient intelligence systems are represented as layer-based multi agent systems that enable the division of system elements into levels, reduce the coupling between modules and facilitates abstractions as well as the distributions of responsibilities. The suggested approach is illustrated with an example from the field of health and medicine, namely the development of Holter monitoring system using context based on O-MaSE methodology and agentTool III.

13.1 Introduction

The Industry 4.0 platform is an initiative of the German Federal Government to support German industry in the transition to digital production with intelligent, digital networks and systems that enable largely self-control and self-management of manufacturing processes [1]. The technological basis of Industry 4.0 is represented by key and core technologies and their interactions. The degree of convergence depends on

T. Ivanova (✉) · I. Batchkova
University of Chemical Technology and Metallurgy, Kliment Ohridski 8 Blvd., Sofia, Bulgaria
e-mail: t.ivanova@uctm.edu

I. Batchkova
e-mail: idilia@uctm.edu

A. Dingli et al. (eds.), *Artificial Intelligence in Industry 4.0*,
Studies in Computational Intelligence 928,
https://doi.org/10.1007/978-3-030-61045-6_13

the field of application and includes a series of buzzwords, such as ubiquitous or pervasive computing, ambient intelligence, cyber-physical systems, context awareness, artificial intelligence and more. The market penetration of more and more diverse embedded and mobile devices set the agenda for solving the problem of ubiquitous computers dealing with that will make possible the successful implementation of diverse and heterogeneous applications in the area of Ambient Intelligence (AmI) [2]. For successful and efficient development of ambient intelligence systems and their components a new approach is needed, aligned to the vision of today's Industry 4.0. The upcoming trends in the development of software systems for the domain of AmI are connected with the well-known Industry 4.0 design principles, summarized in [2], such as interoperability, information transparency, decentralization and autonomy, real-time capability, technical assistance and service orientation (with Human-Machine Interaction) and modularity.

The response to these new challenges is to use new advanced methods in their development life cycle, such as the Model Driven Development (MDD) [4] in order to meet the requirements for reusability, interoperability and cost efficiency. The second very important aspect in the system development is to achieve a reactive and proactive behaviour of the system components through their modelling and implementation as agents. A great number of methodologies, frameworks, meta-models, languages, tools and methods have been proposed to assist and facilitate the development and application of agent-based systems. Unfortunately, however, there are few who have reached the required level of maturity and completeness and can fully respond to the need for agent-based system developments. Other also possible, reasons are the over-involvement of existing methodologies with the agent architecture in the early stages of development and the absence of a clearly defined vocabulary for modelling agent-based systems. The difficulties of implementing multi-agent systems are also widely known, there is a gap between agent—based design and implementation. All these shortcomings and problems can be greatly reduced by the use of a model-driven approach that allows dealing with the complexity and interconnectedness of distributed systems, increasing the agility of software processes, and improving the quality and maintenance of the product created. The main aim of the proposed paper is to suggest an approach for model driven development of multi agent systems for the domain of AmI, based on the combined use of above mentioned advanced approaches of software engineering supported by agentTool III (aT3) tools in Eclipse.

13.2 Related Works

Based on the definition of basic requirements for achieving AmI, a short analysis of related works is represented in this section.

13.2.1 Basic Requirements for Achieving Ambient Intelligence

Ambient Intelligence (AmI) is a vision of Information Society and Technology Advisory Group (ISTAG) [2] for building of smart environments that are reactive and proactive to people and make their actions safer, more efficient, more informed, more comfortable, or simply more enticing. This means that the focus is moved from the individuals that has to adapted, towards of the technologies which need to be adapted to the individuals. As shown in Fig. 13.1, AmI spaces are combinations of two kinds of components; components of operational technologies such as smart materials, I/O devices such as sensors and actuators, ubiquitous communications, adaptive software, etc., and AI components including media management and handling, natural interaction, computational intelligence, context awareness, and emotional computing, emotional computing. The five main features that an AmI system should fulfill with respect to its relation with users can be determined as [5]:

- Non-obtrusive: the system should not interfere with the user's life style through invisible, embedded and distributed devices.
- Context aware: the system should be able to recognize and anticipate the context the user is evolving in and use this context to react to his needs and requirements.
- Personalized: users' profiles and environments are heterogeneous. The system should be customizable in order to fit in different environments and to be used by different users.

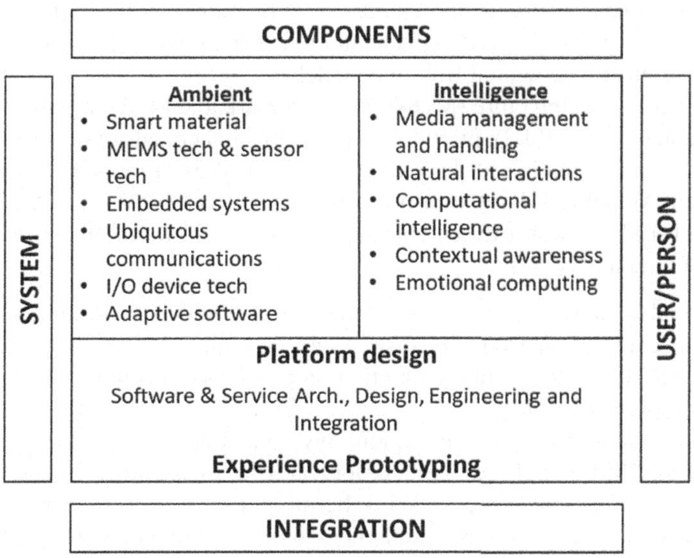

Fig. 13.1 Main research domain supporting AmI according [3]

- Adaptive: the system's behaviour can change in response to the actions and the needs of a person. New services and functionalities can be added with respect to the user needs.
- Anticipatory: it anticipates the person's desires and environment status; it is context predictive and proactively enabler.

These properties of AmI systems cannot be achieved without the use of new methods, technologies and tools, among which stand out the buzzwords pervasive, ubiquitous, context-aware, profiling, etc. Another very important factor in achieving the desired properties of the AmI system is the choice of suitable architecture and software infrastructure that enable easy integration, interoperability and extensibility. Considering these aspects, in this chapter an infrastructure to support an efficient approach for development of AmI systems, combining the approaches of MDD and MAS, is presented. In the next part a short overview and comparison of these approaches is given.

13.2.2 Model Driven Development (MDD)

Model Driven Engineering (MDE) and Model Driven Development (MDD) [4] are some of the most promising and challenging approaches for development and maintenance of highly distributed real time systems such as the AmI systems. Here the systems are presented as models that conform to meta-models, and the model transformations are used to manipulate the various representations. The main difference from other development methods based on models is that MDD uses Models as inputs to parameterized implementation generators, i.e. implementation is (semi)automatically generated from the models.

Model Driven Architecture (MDA) [6] is a remarkable MDD initiative of Object Management Group (OMG), providing a powerful conceptual framework for development and transformation of three interconnected types of models—Computation Independent Model (CIM), Platform Independent Model (PIM) and Platform Specific Model (PSM) towards executable applications as shown in Fig. 13.2. The main features of the models are:

- CIM—is also known as business or domain model that uses vocabulary to present the basic expectation from the system and to bridge the gap between domain experts and developers. This abstraction hides all specifications connected to the system implementation.
- PIM—is a view of the system without any details about implementation. Basic tasks of the PIM model are to model logical data, to establish dependencies and defining workflows and processes. Furthermore, PIM models must be sufficiently complete and accurate to ensure a higher degree of automated implementation of the models in the next layer (PSM).

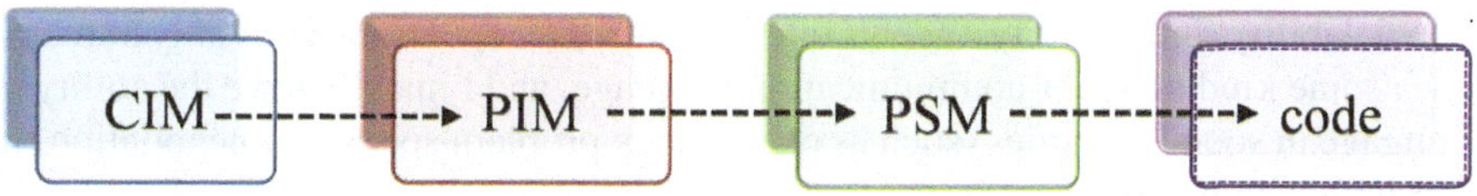

Fig. 13.2 MDA conceptual framework (*Source* Author)

- PSM—combine the PIM specifications with concrete platform information needed for enabling system execution, i.e. the basic role of this model is to ease the code generation using PIM and selected execution platform.

Very important aspect of the MDA approach is the transformation between the different viewpoint models. Model transformation relies on a set of mapping rules between models that inform the transformer tool about the patterns it needs to follow to output the target model [7]. The MDA approach is supported by the OMG techniques like Unified Modelling Language (UML), Query View Transformation Specification (QVT) and Meta Object facility (MOF) [8]. In the core of MDA are the open standards, UML, MOF, XMI, etc. UML [9] provide an integrated modelling framework, covering structural, functional and behaviour descriptions. The UML notations support the development of various diagrams that reflect different aspects of the system in order to capture the full complexity in the phases of detailed analysis and system design.

MDD and MDA are suitable for use in the development of AmI systems and for combination with SOA or MAS, because of the opportunities they create through the applied meta-models and model transformations for providing reusability, verification and validation of models, integration and interoperability. There are some successfully developed agent meta-models and transformation tools (model to model or model to text) for development of MAS, such as for example: SODA, MaCMAS, DMLS4MAS, PIM4Agents, Pineapple, Sol, etc. A review of these models, tools and applications is presented in [10].

13.2.3 Multi Agent Systems (MAS)

Multi Agent Systems (MAS) can be defined as "a loosely coupled network of problem solvers (agents) that work together to solve problems that are beyond the individual capabilities or knowledge of each problem solver" [11]. Jennings and Wooldridge [12] have defined an agent as "a computer system situated in some environment and capable of autonomous action in this environment, in order to meet its design objectives". Agents have the following main properties and characteristics [13]:

- autonomy: agents encapsulate some state (that is not accessible to other agents), and make decisions about what to do, based on this state, without the direct intervention of humans or others;

- socialability (interactivity): agents interact with other agents (and possibly humans) via some kind of agent-communication language, and typically have the ability to engage in social activities (such as cooperative problem solving or negotiation) in order to achieve their goals;
- reactivity: agents are situated in an environment, (which may be the physical world, a user via a graphical user interface, a collection of other agents, the INTERNET, or perhaps many of these combined), are able to perceive this environment (through the use of potentially imperfect sensors), and are able to respond in a timely fashion to changes that occur in it;
- pro-activeness: agents do not simply act in response to their environment, they are able to exhibit goal-directed behaviour by taking the initiative;
- mobility: agents can transport themselves across different systems architectures and platforms.

The agent community has considerable interest in developing methods and techniques for specifying, modelling, implementing and verifying of MAS for distributed information and control systems, as are the AmI systems, but so far no standardized design methodology has been recognized. Several object-oriented methodologies have been suggested for agent-oriented analysis and design, based on UML. Important drawbacks of using UML to model MAS are the modelling of agent communications as method invocations and the absence of references to the mental state of the agents. To overcome these drawbacks, the UML notations are extended to reflect the characteristic properties of the agents. Successfully extensions of UML are achieved in AUML, GAIA, MESSAGE/UML, AgentUML, Prometheus, etc. Some of them are based on FIPA standard (http://www.fipa.org) suggesting an agent reference model for creation, registration, location, communication, migration and retirement of agents. Recently some specialized tools for lightweight devices are also available. Some of them are suitable for the domain of AmI, such as DSML4MAS (http://dsml4mas.sourceforge.net/), FIPA-OS, ASEME, Tropos (http://www.troposproject.org/), INGENIAS (http://sourceforge.net/projects/ingenias/), Jade-Leap, etc. However, there are limitations and drawbacks, associated with the variety of devices and communication protocols, specific for AmI. As well there are some agent-based development environments especially for the AmI domain, as for example: THOMAS, MaRV, ALZ-MAS, CodeBlu, etc. However, they are ad hoc solutions, without considering the high level modelling of AmI properties. There are several different proposals for development of AmI systems and all of them are based on UML. The most popular are UML Profile for context modelling and ContextUML Profile, which are discussed in short below.

13.2.4 Context-Aware Modelling

- UML Context Modelling Profile. The first large subclass of approaches uses context-aware application for AmI systems. A context-aware application for AmI can be modelled using specialized UML diagrams in order to separate concerns and to focus different aspects of the software system. Several works have been taken in the domain of context modelling. For this purpose a special UML Context Modelling Profile (CMP) was created [14]. This profile is a partial extension of a UML meta-model and provides a way to introduce a special terminology for a domain, to add semantics to meta-models and, which is vital, to restrict the way how to use the meta-model. It contains different stereotypes, tagged values and constraints assigned using OCL (Object Constraint Language) defining how an existing meta-class of the meta-model may be extended. The proposed meta-model is then used for modelling context [15]. New meta-classes must not be inserted in the class hierarchy of the meta-model because is forbidden to modify the class definitions in the meta-model. As modelling tool StarUML, which is open source modelling platform in the medical domain, is used. The authors aim to add new stereotypes in the feature in order to be able to model dynamic behaviour of context-aware applications.

- ContextUML profile. ContextUML approach is another UML-based graphical language profile aimed at the design of context-aware web services, which include constructs for modelling context information. It provides constructs for generalizing context provisioning that includes context attributes specification and retrieval and formalizing context awareness mechanisms and their usage in CASs (Context-Aware Services). ContextUML distinguishes between atomic and composite context. However the notion of composite context is different from other approaches which also consider it. ContextUML defines a composite context as a state of a context depending on two or more atomic contexts, and models it by using state-charts [16]. ContextUML models can be expressed in MLContext which is also an MDD approach for the development of context-aware systems. The MLContext meta-model has been designed around the concept of entity. The context of use of a system is represented by defined set of elements that represent all contextual aspects. The type of a contextual element is provided by one of the following entities: the user, the application, the environment or the behavior. So each contextual element has one of the following stereotypes: "UserContext", "ApplicationContext", "EnvironmentContext" and "BehaviourContext". The context information of a simple context can contain a reference to other entities of the model. The information from the source is supplied through a method (Method metaclass) from the physical device interface. The EMF framework of Eclipse as basic tool is used [17]. In ContextUML the acquisition of context information is not modelled. ContextUML covers the functionalities of context-aware services. Modification or change in one element of the context of use involves the transition of this context toward another state of the user's current situation. In order to overcome this modification a specific UML notations and new elements—stereotypes, tagged values and constraints are added [18].

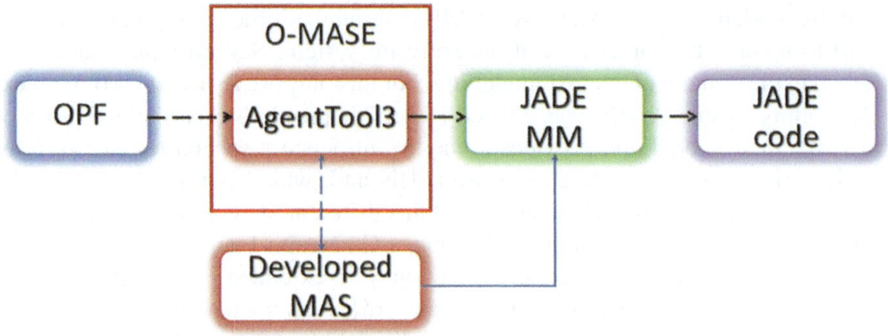

Fig. 13.3 Illustration of the suggested approach (*Source* Author)

13.3 Proposed Work

In order to fulfil the basic requirements to the AmI systems, the software infrastructure of the system is necessary to be distributed, enabling self-organization of devices and their software components. AmI systems are much more complicated than traditional computing systems. Hence, characteristics such as adaptability, flexibility, interoperability and modularity are more important. Furthermore, these systems must provide common improvements such as service discovery, self-organization, rich knowledge representations and context-awareness.

The suggested model driven approach supporting the fulfilment of above mentioned requirements uses MDA and comprises the whole development life cycle of software development, starting with CIM meta-model to the deployment. The approach is illustrated in Fig. 13.3 using 4 layers: one CIM layer, two PIM layers and one PSM layer. Two principles as architecture cornerstones are chosen:

- Model-driven principle using MDA of OMG, supporting the platform independent and model based domain engineering;
- aT3 meta-model [19]—platform-independent model that specifies multi-agent systems in a technology independent manner, based on the O-MaSE (Organization-based Multi-agent System Engineering) process framework [20].

O-MaSE is in form of meta-model, method fragments, and guidelines, based on the OPEN Process Framework (OPF) meta-model in the CIM level of model hierarchy. OPF meta-model is a generic process meta-model defining different types of method fragments and consists of Stages, Work Units (Activities, Tasks, and Techniques), Producers, Work Products, and Languages. The O-MaSE meta-model, shown in Fig. 13.4, looks like an UML class diagram and is based on an organizational approach. The meta-model defines the basic concepts used for definition of multi-agent systems (agent, role, goal, protocol, etc.), represented as classes and the relationships between them using specially defined stereotypes (posses, requires, responds etc.).

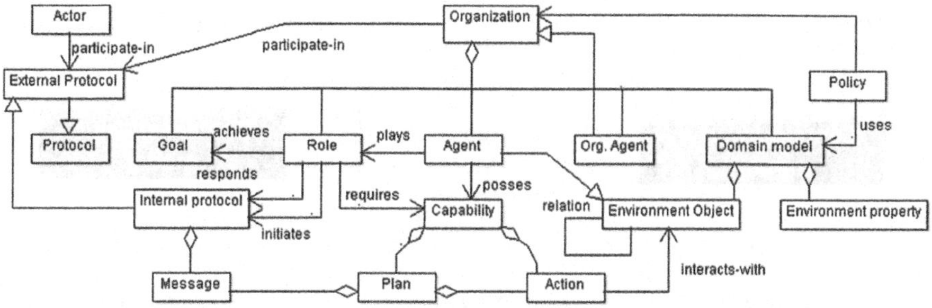

Fig. 13.4 O-MaSE meta-model [20]

The meta-model "AgentTools III" also known as aT3 is used to support the software processes at PIM layer, consisting of four components: graphical editor, process editor, verification framework, and code generation facility, providing abstractions that are tailored to the specific problem domain of agent-based development. The structure of aT3 is divided into several viewpoints each focusing on a particular aspects of MASs. This separation allows extending the core of aT3 in order to reflect additional application-specific requirements. The tool supports an Iterative, Incremental, Parallel Life Cycle model with 4 basic life cycles: requirements engineering, analysis, design, and implementation. The agentTool III (aT3) development environment is built on the Eclipse platform and provides traditional model creation tools to support the analysis, design, and implementation of multi agent systems following the Organization-based Multi-agent Systems Engineering (O-MaSE) methodology. It also provides the ability to compose, verify, and maintain customized O-MaSE complaint processes. Additionally, aT3 provides a verification framework which helps designers to maintain consistency between their OMaSE models. AgentTool III is a suite of Eclipse plug-ins that facilitate the design and analysis of multi agent systems. Currently, agentTool III provides the following eight diagramming plug-ins: Agent Diagram, Capability Diagram, Domain Diagram, Goal Diagram, Organization Diagram, Plan Diagram, Protocol Diagram and Role Diagram. With the proposed approach it is possible to design an agent-based AmI system specifying high level concepts in a platform-independent agent model (focusing on the domain model), and later automatically transform it for different implementation models, bridging the gap between design and implementation. The use of two layers of PIM models enables for horizontal transformation between the models of the second layer and the use of "foreign" environments for performance.

An extension to the traditional MDD approach is the suggestion for use of compositional models as core elements for automation of expert activities, enabling the context-awareness of application. Compositional models encapsulate knowledge on a specific problem and are modeled in two dimensions—domain and context. Such a way the domain specific concepts have meaning only in relation to contextual concepts. Whereas the domain-specific models are formal descriptions of the domain concerned, the context-awareness models are representations of information related

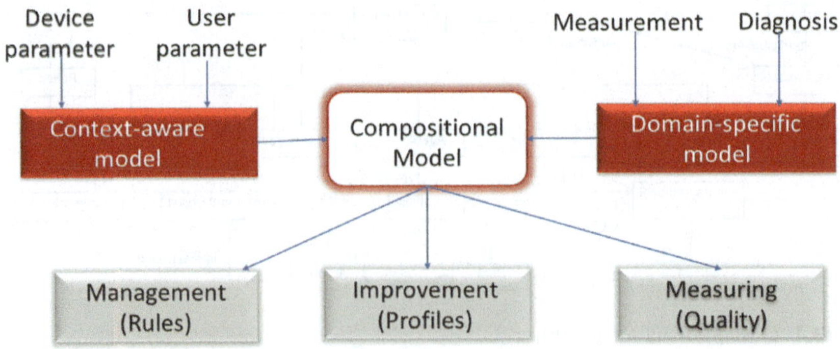

Fig. 13.5 Use of compositional models (*Source* Author)

to the domain environment. As is evident from Fig. 13.5 the domain-specific model uses measurements and diagnosis results based on measurements, while the context-awareness model is based on device and user parameters. The compositional model may be seen as a collection of data, rules, and terminology leading to transparent understanding of a particular process. The process implementation is supported by management, measuring and improvement procedures. The terminology of the compositional model allows describing user profiles, which may be aggregated in generic or reference profiles participating in quality assessments and process improvements.

13.4 Case Study (Discussion)

In this section the suggested approach is illustrated with a case study from the domain of healthcare.

13.4.1 Short Description of Holter Monitoring Systems

The above suggested approach will be demonstrated with a case study from the field of healthcare. It is about using telemedicine to prevent and improve human health through monitoring systems. Many heart problems become noticeable only during different type of everyday life activities, and electrocardiography (ECG), also known as Holter monitoring, has been used in order to support diagnostics and to prevent unpleasant incidents. In recent years, Holter monitoring has expanded its application and is now being used for diagnosis of cardiac arrhythmias, quantification of arrhythmias, evaluation of the antiarrhythmic therapy, detecting the pacemaker malfunctions, in heart rate variability studies, and in the prognostic stratification of arrhythmias [21]. It has also been used in detecting the silent episodes of

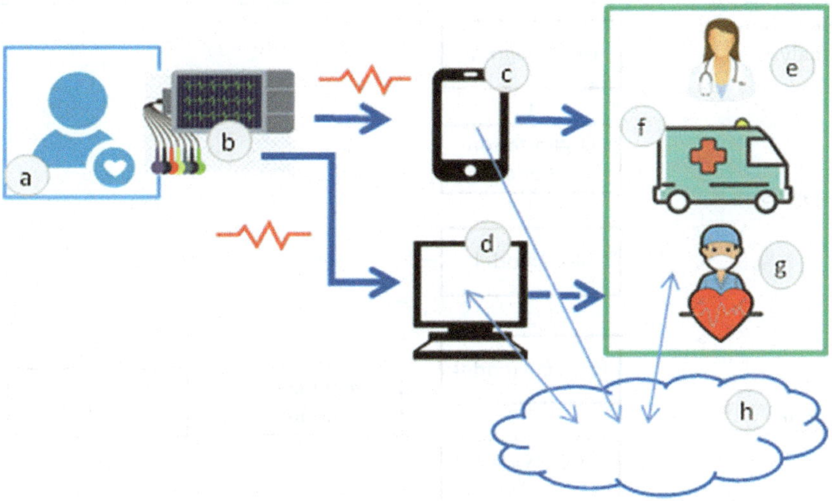

Fig. 13.6 Proposed Holter monitoring system (*Source* Author)

myocardial ischemia and in the evaluation of various nonspecific symptoms like palpitations, light-headedness, blackouts, drop attacks and atypical chest pain. These facts impose greater demands on data processing in monitoring systems, expansion of the communications and interactions in the system and the possibility of using a variety of computational tools for diagnostic purposes.

For the purposes of Holter monitoring, a patient wears a small recorder called a Holter monitor as he or she goes about normal daily life. Holter monitoring system not only detects the presence of abnormal heartbeat intervals but also requests for context information about the activities undertaken by the patient upon detecting changed heart rate [22]. The main components of Holter monitoring system are shown in Fig. 13.6 and include Holter with Bluetooth (b) and Personal Cardio Assistant (PCA) for ECG beat detection. There is a possibility a smart phone to collect the information of received ECG signals (c) and by patient or doctor's advice, this information can be recorded and stored on a local home computer (d) for more detailed analysis. The so suggested architecture of Holter monitoring based on PCA must provide a high degree of mobility and appropriate agent interactions to ensure preventive monitoring of heart activity while the patient is either in stationary or mobile situation.

13.4.2 Model-Driven Development of Agent Based Monitoring System

The development of agent based monitoring system follows the software model (development life cycle model) shown in Fig. 13.7 and includes 4 phases (require-

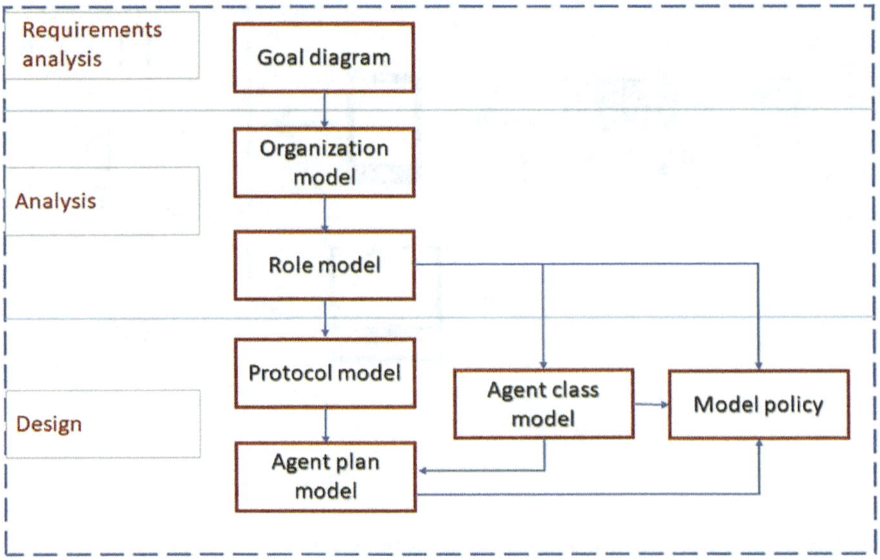

Fig. 13.7 Suggested software process model (*Source* Author)

ments definition, analysis, design and implementation). The last phase of the development life cycle is not illustrated on the figure.

The nowadays Holter monitoring systems can send data in real time to a specialized call centre where the doctors take decisions according to the received data. Patients should keep a record of their activities during the whole period. Taking into consideration the advanced development and capabilities of the information and communication technologies as well as the progress in the field of medical equipment and devices, within the developed project, the following requirements for the Holter monitoring system are defined:

- The system should gather patient data and keep a track of the patient's activities (like sport, rest, sleep, etc.). Based on these records and on the pre-defined ECG annotations, pre-programmed decisions should be made;
- Evaluate patient data and store the results;
- The Holter monitoring system should not only send information to call (emergency) centres but should also take decision on its own.

The development processes start with the phase of requirements definition and modelling the above defined functional requirements through the goal diagram, based on tree structure and AND/OR relationships for defining the selected functional and non-functional requirements to the project. Trigger relationships may also be used for defining dynamical appearing goals. The goal model is shown in Fig. 13.8 and includes 3 main goals described above. The main goals are refined by the definition of sub-goals. At this stage of development 10 sub-goals are defined.

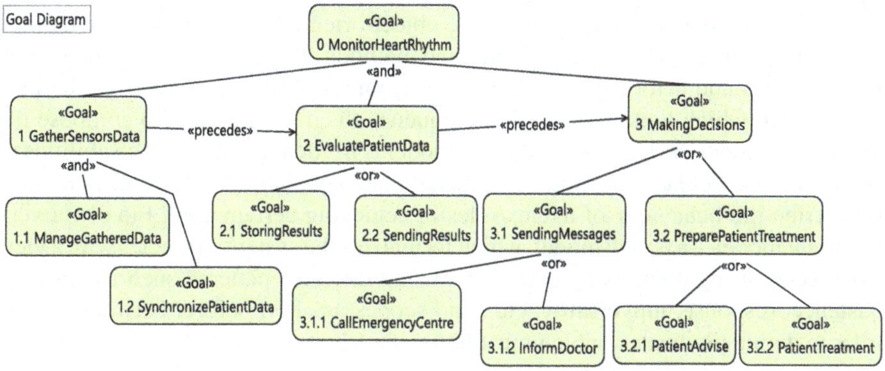

Fig. 13.8 Requirement definitions through Goal diagram (*Source* Author)

The goal model is followed by the analysis phase and the creation of static diagrams such as organization model and role diagram, which are similar to the UML class diagram. The organization model represents the relationships of the system to external users. The Role Model defines the type of roles in the organization and the goals they are designed to achieve. This model supports the creation of all models at the design phase, such as protocol diagram, agent class model and model policy. The agent class model defines the agents used in the project and their relationships with actors, organizations, roles, capabilities and protocols. The developed agent class model is shown in Fig. 13.9 and includes 4 agents characterized by defined capa-

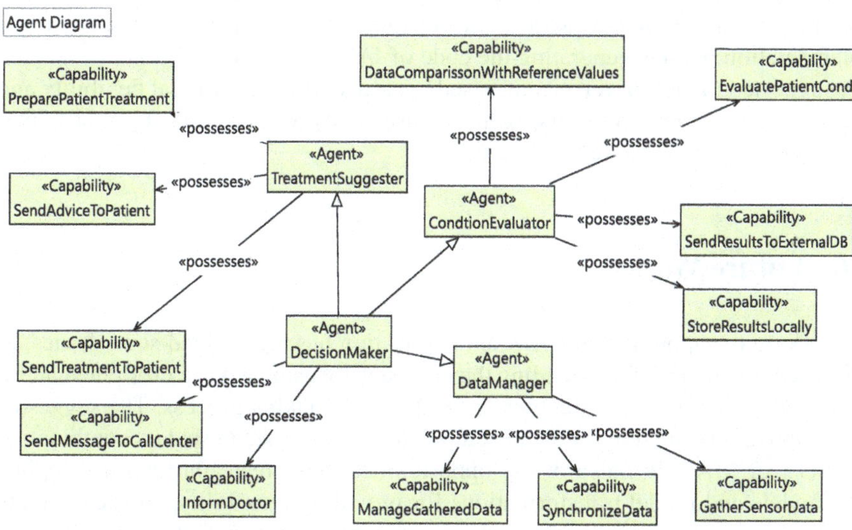

Fig. 13.9 Agent class diagram (*Source* Author)

bilities. The agent classes differ from the object-oriented, because they are defined through roles, instead of using attributes and methods. The interactions in the system between roles and actors are described as sequences of messages through the Protocol model, which look like an UML Sequence diagram. In order to compose the Agent Plan model the Capability-Action model is needed that defines the abilities of agents or roles, defined as actions. The Agent Plan model uses Finite State Automata to describe the behaviour of agents/roles in achieving certain goal. Finally, a very important model—a policy model, which formalizes the system's policies with a view to verifying it, by composing quantifiers using different patterns such as absence, existence, response, universality, etc., and scopes as after, before, between, etc. in order to define predicates, which must be verified.

13.5 Conclusion

The approach proposed in this paper attempts to cope with the requirement to achieve the context-aware of AmI applications using the model driven architecture and compositional models. The suggested approach uses the MDA to address the modelling of MAS in order to facilitate software development through abstractions avoiding the specific implementation details. The main idea is to achieve a layer based design that enables the division of system elements into levels, reducing the coupling between modules, facilitating abstraction as well as the distribution of responsibilities. The choice of PIM meta-models is essential to achieve the requirement for context awareness of AmI systems. Some of the main reasons to choose the O-MaSE methodology and process framework, and the agentTool III meta-model are the possibilities to customize the development process and to fully fill the gap between design and implementation phases, generating the code of JADE based application.

A suggested model-driven, agent-based approach offers additional flexibility and advantages (in general and in the scenario discussed) when agents are tightly integrated into a service-oriented framework.

13.6 Future Work

The work on the approach and selected application has highlighted some issues, as well as generated ideas for extending the approach by using a service-oriented architecture and incorporating ontologies in the decision-making process. The suggested MDD approach is suitable for the development of AmI systems and its combination with SOA or MAS, because of the opportunities they create through the applied meta-models and model transformations for providing reusability, verification and validation of models, integration and interoperability. An important step forward would be to increase the degree of adaptability by incorporating elements of machine learning.

Acknowledgements This work is financially supported by the National Research Fund of the Bulgarian Ministry of Education and Science in the frame of project KN-06-H27-8/2018.

References

1. Kagermann, H., Wahlster, W., Helbig, J.: Recommendations for implementing the strategic initiative INDUSTRIE 4.0, Final report of the Industrie 4.0 Working Group, Akatech, München, pp. 19–26 (2013)
2. IST Advisory Group: Ambient Intelligence: from vision to reality. In: Riva, G., Vatalaro, F., Davide, F., Alcañiz, M. (eds.) Ambient Intelligence, pp. 45–68. IOS Press (2005)
3. i-SCOOP: Industry 4.0: the fourth industrial revolution – guide to Industrie 4.0 (2019). https://www.i-scoop.eu/industry-4-0/. Accessed 15 March 2019
4. Kent, S.: Model driven engineering. In: Butler, M., Petre, L., Sere, K. (eds) IFM 2002, Turku, Finland, LNCS, vol. 2335, p. 286. Springer, Heidelberg (2002)
5. Aarts, E., Marzano, S.: The New Everyday: Visions of Ambient Intelligence. 010 Publishers, The Netherlands, Rotterda (2003)
6. Object Management Group OMG-MDA: MDA Guide version 1.0.1 OMG document omg/2003-06-01 (2019). https://www.omg.org/news/meetings/workshops/UML_2003_Manual/00-2_MDA_Guide_v1.0.1.pdf. Accessed 15 March 2019
7. Gorp, P.: Model-driven development of model transformations. In: Ehrig, H., Heckel, R., Rozenberg, G., Taentzer, G. (eds) Graph Transformations. ICGT, Lecture Notes in Computer Science, vol. 5214. Springer, Berlin, Heidelberg (2008)
8. Object Management Group: About the meta object facility specification 2.5.1 (2019). https://www.omg.org/spec/MOF. Accessed 15 March 2019
9. Object Management Group: About the unified modeling language specification 2.5.1 (2019). https://www.omg.org/spec/UML/About-UML/. Accessed 15 March 2019
10. Vinas, I.A.: Model Driven Development of Agents for Ambient Intelligence. Dissertation, Universidad de Malaga (2013)
11. Durfee, E.H., Lesser, V.: Negotiating task decomposition and allocation using partial global planning, In: Gasser, L., Huhns, M. (eds) Distributed Artificial Intelligence, vol. II, pp. 229–244. Pitman Publishing, London, Morgan Kaufmann, CA San Mateo (1989)
12. Jennings, N.R., Wooldridge, M.: Applications of agent technology. In: Jennings, N.R., Wooldridge, M. (eds.) Agent Technology: Foundations, Applications, and Markets. Springer, Heidelberg (1998)
13. Wooldridge, M., Jennings, N.R., Kinny, D.: The gaia methodology for agent-oriented analysis and design. Int. J. Auton. Agents Multi-Agent Syst. **3**(3), 285–312 (2000)
14. Simson, Ch.: CMP: A UML context modeling profile for mobile distributed systems. In: Proceedings of the 40th Hawaii International Conference on System Sciences, Hawaii (2007)
15. Benselim, M.S., Seridi-Bouchelaghem, H.: Extending UML class diagram notation for the development of context-aware applications. J. Emerg. Technol. Web Intell. **5**(1), 35–44 (2013)
16. Sheng, Q.Z., Benatallah, B.: ContextUML: A UML-based modelling language for model-driven development of context-aware web services. In: Proceedings of 2005 International Conference on Mobile Business (ICMB 2005), Sydney, Australia, pp. 206–212 (2005). Accessed 11–13 July 2005
17. Hoyos, J.R., Garcia-Molina, J., Botıa, J.A.: MLContext: a context-modelling language for context-aware systems, In: Proceedings of the Third International DisCoTec Workshop on Context-Aware Adaptation Mechanisms for Pervasive and Ubiquitous Services (CAMPUS 2010), vol. 28, Amsterdam, Netherlands (2010)
18. Benselim, M.S., Seridi-Bouchelaghem, H.: Modelling context with extended UML. AWER Proc. Inf. Technol. Comput. Sci. **1**, 566–571 (2012)

19. Garca-Ojeda, J.C., DeLoach, S.A., Robby: AgentTool III: From process definition to code generation. In: Decker, S., Sierra, C. (eds) Proceedings of 8th International Conference on Autonomous Agents and Multiagent Systems (AAMAS), pp. 1393–1394, Budapest, Hungary (2009). Accessed 10–15 May 2009
20. DeLoach, S.A.: Engineering organization-based multi-agent systems. In: Garcia, A., Choren, R., Lucena, C., Giorgini, P., Holvoet, T., Romanovsky, A. (eds.) Software Engineering for Multi-agent Systems IV: Research Issues and Practical Applications, pp. 109–125. Springer, Berlin (2006)
21. Paudel, B., Paudel, K.: The diagnostic significance of the holter monitoring in the evaluation of palpitation. J. Clin. Diagn. Res. **7**(3), 480–483 (2013)
22. Li, J.: A mobile ECG monitoring system with context collection. Dissertation, Dublin Institute of Technology (2008)

Chapter 14
Designing Trust in Highly Automated Virtual Assistants: A Taxonomy of Levels of Autonomy

Fernando Galdon, Ashley Hall, and Stephen Jia Wang

Abstract This paper presents a guiding framework and a multi-level taxonomy of automation levels specially adapted to Virtual Assistants in the context of Human-Human-Interaction. This trust-based framework incorporates interaction phases, trust-affecting design principles and design techniques. It also introduces a taxonomy of levels of autonomy explaining each level from a trust perspective. To test the proposed Levels a survey was conducted addressing different contexts. Participants preferred to have total control of the system. Level 1 is the preferred option on average. Levels 2 and 3, account for 40.50% of the participants preference to be in control of the autonomous system. If we combine levels 1, 2, and 3; This presents an average of 68.75% of participants demanding the initiative. The neutral level (level 4) is preferred by 15.75% of the participants on average. On Levels where the initiative resides on the system (levels 5, 6, and 7), only 14.75% of participants would decentralise their decision. Based on the research findings, the author recommends designers to combine a holistic perspective on trust with contextual awareness, to be able to integrate the impact of contexts on interactions. Trust formation is a dynamic process that starts before a user's first contact with the system and continues long thereafter. Furthermore, as autonomous systems continuously evolve, factors-affecting trust change during user interactions with the system and over time; thus, Human-Human-Interaction concepts need to be able to adapt. Future work will be dedicated to further understanding other areas such as reparation and accountability.

F. Galdon (✉) · A. Hall · S. J. Wang
Royal College of Art, London, UK
e-mail: fernando.galdon@network.rca.ac.uk

A. Hall
e-mail: ashley.hall@rca.ac.uk

S. J. Wang
e-mail: stephen.wang@rca.ac.uk

A. Dingli et al. (eds.), *Artificial Intelligence in Industry 4.0*,
Studies in Computational Intelligence 928,
https://doi.org/10.1007/978-3-030-61045-6_14

14.1 Introduction

Recently, Peter Hancock presented a warning to the field of human factors in which attention must be given to the appropriate design of a new class of technology: highly autonomous systems [20]. In the context of using such an autonomous system as Virtual Assistants (VA), this warning has become factual with a demo presented by Google called Duplex. This VA system demo presented an extraordinary level of fluidity, coherence, and autonomy never seen before.

With the development and combination of machine learning and deep learning techniques, a new paradigm is raising: Machine-Human interaction (MHI). In this paradigm the technology is holding the initiative of the communication. This approach positions highly autonomous systems at the centre and tries to address the implications of trust from their perspective [31]. However, the nascent nature of these systems and the unavailability of them to conduct research prevents an adequate development of strategies. Traditionally, complex autonomous systems required the human operator to appropriately calibrate their trust in the automation to achieve performance and safety goals. In this context, literature has focused on the human-machine- interaction (HMI), and human-human-interaction (HHI) trust paradigms to precisely define and measure trust in automation. In this article, the author minds the warning and propose a human centred approach in the context of HHI directly aimed at ensuring that emerging highly autonomous systems interactions remain focused on the user's needs and preferences in the context of Virtual Assistants.

14.1.1 From Automation to Autonomy

Automation is defined as a system with a limited set of pre-programmed supervised tasks on behalf of the user. Autonomy, on the other hand, is defined as a technology designed to carry out a user's goals without supervision with the capability of learning and changing over time. These systems can dynamically set its own goals, holds the ability to adapt to local conditions via external info (sensors/input) and have the potential to evolve in unexpected ways [27]. Because of the increasing unpredictability of these systems, they may also be more likely to surprise human partners to an even greater extent than simple automated systems [49]. It is precisely this unexpected nature what raises concerns for trust in these emerging highly automated systems.

As we are placing this debate in the context of virtual assistants, We can observe a clear distinction by conducting a comparative study among Alexa and Duplex. The usability of Alexa is based on a one-off query [53]. The system has the ability to stream audio over Bluetooth, request radio stations, play music, make lists, ask about the weather and news, and order products from https://www.amazon.com/ [53]. On the other hand, Duplex is a system presenting an extraordinary level of fluidity, coherence, and autonomy never seen before. In a demo introduced by Google in May

2018, the system was able to make a hair appointment with a human without any supervision. This system, not only was able to deliver the task but did it without the human noticing that she was speaking to a robot. This system differs radically from Alexa in the sense that we are moving from one-off queries to conversations. And that the initiative in the interaction is not necessarily placed in the user but within the system.

Recent examples of unexpected behavior in highly automated systems are for instance Libratus. This system was designed to play poker. It used a combination of deep learning and machine learning. The surprise came when the researchers reviewed the data and found out that the system had learned to lie. Not only learned to lie but it implemented the strategy in the right moments [4]. In another experiment conducted by Facebook, the researchers confronted two chat-bots to improve selling strategies. The surprise came when the researchers went back and discover that the systems developed a language to communicate between them. As with the case before presented, the designers did not design the system to do it [19]. In another case, an AI agent developed by Microsoft and the University of Cambridge, Deep-Coder, took lines of code from existing software looting other software in the process. This system has learned how to write programs without prior knowledge of code [2]. Finally, in the context of virtual assistants, Alexa recorded a private conversation and sent it to other users without the main user knowing [9]. The unpredictability on how these unsupervised agents may evolve and its goal-oriented nature present a fundamental problem for society and businesses. Is this regard a fundamental question arises; what strategies would enable us to establish and maintain trust in these systems?

14.1.2 From Machine-Human to Human-Human

Three main paradigms emerge on how to approach trust in Highly Automated Systems; human- machine-Interaction (HMI), human-human-Interaction (HHI) and an emerging machine-human- interaction perspective (MHI).

From an HMI perspective Interpersonal trust conceptualizations only provide a limited explanation. Distinct psychological constructs and mechanisms need to be considered to explain Human–Agent trust. In this paradigm, Anthropomorphised agents are related to lower initial trust. Computer systems are believed to be more capable, rational and objective than humans. (Automation bias: authority hypothesis) [12, 33, 39, 40].

From an HHI perspective the same psychological constructs and mechanisms can be applied to explain interpersonal and Human–Agent trust. In this paradigm, Computers are social actors (media-equation hypothesis). Form a design perspective, Anthropomorphised agents are related to higher initial trust because Anthropomorphism makes novel systems more familiar and controllable [3, 7, 42, 47].

Finally, in a recent article posted in Medium by the DeepMind safety research team, they advocated focusing on a Machine-Human interaction (MHI) perspective. Their approach implies a superiority of the system to the agent. This strategy is focused on the autonomous system and design strategies are focused on how to control the system [31].

Although this emerging MHI perspective will be the dominant paradigm in the future, several reasons prevent us from adopting this paradigm nowadays; the problems of interpretability (the system does not know what is doing) and interruptibility (the system knowing when to stop) in addition to the nascent nature of this systems and the unavailability of them to conduct research prevent researchers to adopt this paradigm. In terms of HMI perspective, this was the paradigm used until 2017. The fundamental reason for the evolution to an HHI perspective is the nature of neural networks capable of learning form users interactions. In this context, de Visser proposes that we should approach the design of highly autonomous system from an HHI perspective by moving from focusing on interactions to focusing on relationships [11]. Consequently, In this paper, the authors positions the investigation from an HHI perspective.

14.2 Factors for Designing Trust in Automation

Current models structure trust in automation from three different perspectives; the user, the environment and the system [21, 22, 30, 50]. In this context, we can synthesize three main interactive stages; expectations; experimentation and reliability.

1. Expectations depend on preliminary knowledge, recommendation by relatives or endorsement by celebrities and appearance design attributes such as typological design or name.
2. Experimentation is focused on design attributes such as communication style, ease-of- use or transparency-feedback. Elements such as pitch and porosity, intonation and wording or the system sounding comfortable and natural define its communication style. Fluidity and autonomy, the uselessness of recommendations and low error-rates define its ease-of-use and transparency and the communication of intend define its transparency and feedback level.
3. Reliability focuses on design strategies for reducing error-rates within an automation system fundamentally based on stages and levels of automation build around calibration systems.

Research in the area of human factors present evidences that the more reliable the system, the more likely is to be trusted [26, 37, 43–45, 48]. Therefore, positioning this area as the most relevant for building and establishing trust in HAS. In this context, as far as the error-rate is around 30% or less, users will continue using the automated system [46, 55, 57].

In terms of interaction, when users interact with automated systems misuse and disuse are the most common outcomes [13]. When the system fails less than 30% it leads to user's misuse. Misuse refers to cases when the automation is used without skepticism, tending to result in overuse [1, 43, 44]. The main implications are;

automation bias and complacency [35, 43]. They fundamentally arise due to a lack of monitoring where lack of attention plays a central role [43].

On the other hand, when the system fails more than 30% it leads to user's disuse. Disuse relates to an interaction that extends from the user barely using the automation to completely abandoning it in favour of a manual approach [13]. The main reasons are a high expectation of automation performance and unexpected errors [10, 28, 38, 44].

Finally, Hoff, K. A., and Bashir, M. presented in a seminal paper the five fundamental principles to account when designing trust in automation; appearance (P1), Communication style (P2), ease of use (P3), transparency and/or feedback (P4) and Levels of control (P5) [22].

14.2.1 Framework

With these element we can build a framework (Fig. 14.1) specifically adapted to Virtual Assistants to understand the dynamics of the system.

14.2.2 Two Challenges: Reliability and Predictability

As autonomous systems become more and more complex, the ability of users to understand the system becomes more difficult. The higher the levels of automation, the more complex and less understandable the system, the lower the levels of trust [32]. In this context, reliability and predictability have been identified as a key factor influencing trust in automation [30]. Therefore, in order to address trust in highly automated systems, Trust must be appropriately calibrated to the actual system performance [41]. In the context of reliability, predictability has been identified as a fundamental quality for trust in automated systems. It is argued that prediction is necessary to mitigate potentially detrimental interaction behaviour to avoid unwanted results which may result in situations that cannot be changed [13].

In this context, for the system to enhance reliability, the calibration system must enhance predictability. In Predictability, prior knowledge about potential automation failures reduces the level of uncertainty and risk [32]. Once reliability has been judged, the most important factor of trust in automation is predictability of performance over time [29]. Predictability is enhanced by implementing levels of automation. The idea of gradient-base models of approximation with positive, negative and neutral spectrum has been embodied through the concept of scales or Level of trust (LoT). The notion of different levels of automation has been persistent in the automation literature since its introduction by Sheridan and Verplanck [51]. Kaber [25] points out that levels of automation (LoA) is a fundamental design characteristic that determines the ability of operators to provide effective oversight and interaction with system autonomy. Levels aim to improve transparency by simplifying inter-

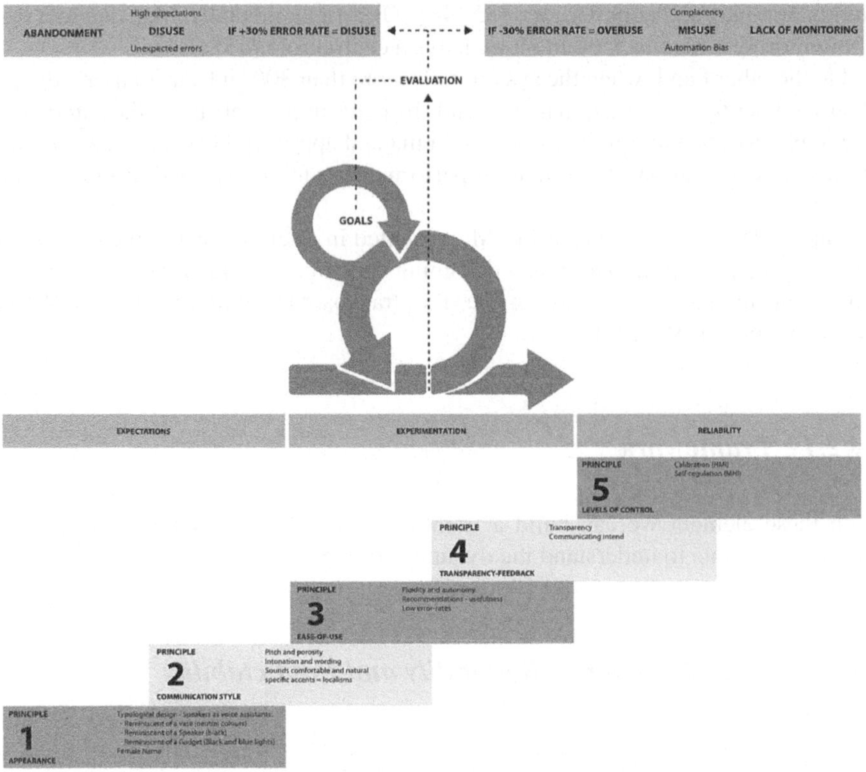

Fig. 14.1 Framework of interactive phases on virtual assistants. It includes design principles (presented by [22]), Phases of interaction, and interactive engagement rates (*Source* Author)

actions. In this context, transparency refers to the extent to which the actions of the automation are understandable and predictable [14]. Automated systems which clarify their reasoning are more likely to be trusted [22, 52].

In this context, prominent frameworks in the area of levels of automation (LoA) are for instance [46]. These researches present a framework which differs radically from past approaches. Instead if structuring a scale, they propose a 4 levels structure outlining four classes or types of automatons functions to account in human-machine- interaction. Wickens et al. [58] degrees of automation proposes a similar approach to Parasuraman with a small addition of the notion of degrees (high and low). An approach more related to [51] is presented by [56]. They present a seven-point scale ranging from total human control to total autonomy in the context of air traffic management. Trimble et al. [54] Multi-variable framework integrates both approaches; stages and levels. This framework is widely used in current vehicle autonomy research. Finally, the most recent taxonomy is presented by [24]. This framework shifts the priority from levels of automation to levels of control given a particular situation. It introduces flexibility and contextual awareness.

As we have reviewed, a range of models of levels of automation (LoA) exits in the area, however, no scale has been designed specifically to address Virtual Assistants in its developmental nature leading to an increasing level of autonomy. This paper will present the first level of automation designed ad-hoc to address present and near-future developments of these systems towards increased autonomy.

14.3 Method

Scales range from one to ten points. The most common types are odd or uneven scales which allow the participant to record a neutral trust level. The most commonly used validated scale is developed by [23]. It is a seven-point scale articulated to measure global trust in automation. Recent studies using the scale presented excellent internal reliability (Cronbach alpha = 0.93) [6]. Other scales include; [36] propensity to trust scale, [29] subjective rating scale, the Human-Computer Trust questionnaire by [34], and a cross-cultural trust in automation scale by [8]. Their functionality for measuring trust ranges from particular types of automation, such as autonomous vehicles [18], to robotics [59].

However [5], argue that the notion of levels of automation are problematic because;

1. Levels reinforce the notion that automation and human activities are interchangeable.
2. It is not clear that a given action is 'higher' or lower' than another action. The Level of autonomy is relative to the task, goals, and context.
3. They promote reductive thinking, for example, considering simultaneous activities as sequential activities.
4. LoA are limited to address current and future challenges.
5. The concept is too human-centred.

Endsley [56] counter-arguments that the most important benefit in LoA is its communicative value to key stakeholders (e.g., system operators, designers, and program managers) about the intrinsic notion that there are different ways and degrees to implement automation. The fact that there is a whole range of options between fully manual and fully automated enhances the understanding of these systems by non-experts. This method has proven successful in providing a solid foundation to understand HAI at a deeper level. This is highly relevant when confronting an invisible entity making decisions while working in the background.

In this context, [25] points out, the LoA decision and design must be made by the system developers. Including adaptive automation, a granularity of control, and automation interface design, LoA is a fundamental design characteristic that determines the ability of operators to provide effective oversight and interaction with system autonomy. In this context, LoA remains a central design decision associated with the design of automated and autonomous systems that must be addressed in future system design.

LEVEL 1	NO AUTONOMY	The VA does not implement the action unless requested by the user
LEVEL 2	ASSISTANCE	The VA assist determining a range of options related to user's query.
LEVEL 3	PARTIAL AUTONOMY	The VA engage in conversation and suggests one option.
LEVEL 4	CONDITIONAL AUTONOMY	The VA selects action and implements it if human approves.
LEVEL 5	RELATIONAL AUTONOMY	The VA selects action, informs human with plenty of time to stop.
LEVEL 6	HIGH AUTONOMY	The VA can perform decisions solely on its own and necessarily tells human what it did
LEVEL 7	FULL AUTONOMY	The VA can perform decisions solely on its own without reporting to the user.

Fig. 14.2 Proposed levels of autonomy for highly automated virtual assistants (*Source* Author)

14.3.1 Designing Levels of Automation for Highly Automated Virtual Assistants

Building from these insights, this study proposes the articulation of an odd scale. This type of scale proposes a neutral element and two extremes which allocate extreme perspectives. In this case no autonomy and full autonomy. As we have seen in the section before, taxonomies are structured between five to ten point scales. However, ten point scales overlap intentionalities in automation. Whereas five point scales seem too narrow to cover the range of possible interactions. They only provide a step in between the neutral and the end extreme measures. In this context, and following these insights, this study proposes a seven point scale which allocates two steps in-between. It uses [51] as its foundation to adapt the scale to Virtual Assistants and the increasing level of autonomy that is expected to evolve in future developments.

14.3.1.1 Proposed Levels of Automation

See Fig. 14.2.

14.3.2 Discussion

Kaber [25] places the responsibility to design the scale Level of Automation (LoA) in the designer. Due to the highly contextual nature of virtual assistants, a preliminary investigation underpinned four highly sensitives areas where highly automated virtual assistants may impact significantly users; health and well-being, identity, economically related activities and social interactions.

Once the relevant contexts were identified, a workshop was conducted with design students at the Royal college of Art to map unintended consequences in these highly sensitive areas;

1. health—death, harm, injury, dependency, addiction.
2. Social—lack of diversity, privacy, dependency, accuracy, stalkers, hackers.

3. Identity—segregation, isolation, manipulation, homogeneity, dependency.
4. Economics—homogeneity, dependency, manipulation, lack of diversity, indoctrination, unethical investments.

From this activity four main categories of unintended consequences emerged;

1. unhappiness about the service—unethical investments, indoctrination, manipulation, addiction.
2. wrong predictions—accuracy.
3. Losing something—dependency, privacy (stalkers, hackers), segregation, isolation, addiction, indoctrination, manipulation, homogeneity, lack of diversity.
4. ends violently—death, harm, injury, addiction.

From the areas aforementioned and based on demos, patents and prototypes, eight cases study were built to address different outcomes. Two cases here build to address each sensitive area ranging from low to high impact. Then, a survey was designed to establish whether the proposed levels of autonomy in highly automated virtual assistants were sufficient to address all the cases.

The primary technique consisted on integrating an 'other' tab to test the scale. This space allowed participants to propose a missing new level (or area). The survey was structured around three sections addressing different aspects of trust. Each section contained the eight cases with their correspondent levels and the 'other' tab. One of the sections was designed to enhance an 'other' output by limiting the levels from seven to four.

50 participant, 21 men, 27 women and 2 who did not want to identify themselves, from 14 different countries with an age range between 18–67 years old from different professions have undertaken the survey and neither of them proposed a new level of automation. Besides, the co- design activity generated 24 variations of unintended consequences and all of them could be covered by the framework proposed.

The survey aimed to understand whether or not contexts affected the level of automation. In the central area of levels of autonomy, contexts play a role in determining which level of automation was needed, but it did not play any role in determining the spectrum. A generic scale covering from no automation to full automation is capable of addressing different contexts in highly automated virtual assistants. However, contexts determine which is the most appropriate level.

44% Participants engaged with the 'other' tab through the survey at different points. Whereas 66% of the participants engaged with the proposed levels as they were presented.

14.4 Conclusions

In the central area of levels of autonomy, contexts play a role in determining which level of automation was needed. However, it did not play any role in determining the spectrum. A generic scale covering from no automation to full automation is capable

AUTONOMY

AREA	PREDICTING cough	PREDICTING depression	PREDICTING politics	PREDICTING sexuality	PREDICTING investment	PREDICTING best job	PREDICTING dating partner	PREDICTING domestic violence	TOTAL
LEVEL 1 No autonomy	24%	20%	**48%**	**60%**	**16%**	10%	**38%**	10%	**28.25%**
LEVEL 2 Assistance	24%	24%	16%	20%	12%	**32%**	16%	14%	**19.75%**
LEVEL 3 Partial autonomy	**26%**	**30%**	20%	8%	22%	18%	26%	**16%**	**20.75%**
LEVEL 4 Conditional autonomy	18%	10%	6%	4%	**34%**	28%	12%	14%	**15.75%**
LEVEL 5 Relational autonomy	6%	14%	8%	8%	8%	6%	2%	**16%**	**8.50%**
LEVEL 6 High autonomy	2%	2%	2%	0%	4%	2%	0%	**16%**	**3.50%**
LEVEL 7 Full autonomy	0%	0%	0%	0%	4%	2%	4%	12%	**2.75%**
OTHER ...	0%	0%	0%	0%	0%	2%	2%	2%	**0.75%**

Fig. 14.3 Survey results (*Source* Author)

of addressing different contexts in highly automated virtual assistants. However, contexts determine which is the most appropriate level (Fig. 14.3).

In terms of contexts affecting levels of trust in autonomy, the most susceptible context for users was Identity with a 54% blended average; the system being capable of predicting sexual or political orientation led to 60% of participants in the first case and 48% in the second to opt for Level 1 (No autonomy). Second most susceptible contexts were Health and well being and social interactions with 22% and 24% blended average respectively. However, the most distributed result was on the system predicting domestic violence. This is the only context where more than 10% of the participants would use high or full autonomy.

Participants preferred to have total control of the system. Level 1 is the preferred option on average. Levels 2 and 3, account for 40.50% of the participants preference to be in control of the autonomous system. If we combine levels 1, 2, and 3; This presents an average of 68.75% of participants demanding the initiative. The neutral level (level 4) is preferred by 15.75% of the participants on average. On Levels where the initiative resides on the system (levels 5, 6, and 7), only 14.75% of participants would decentralise their decision. Domestic violence is the only context where more than 10% of participants would allow the initiative in the system on levels 5, 6, 7. Depression is the other area where participants would go beyond 10%, but only in level 5.

Although several models address the nature and practice of automation systems, models in automation leading to autonomy specifically designed for Virtual assistants focusing on trust remained unsolved. The model presented is a first step in building a system capable of building and maintaining trust in Highly Automated Systems

(HAS), however, other areas such as reparation and accountability must be further investigated.

In this context, two more papers have been published addressing these variables [15, 16], and a final paper integrating this triangulation in an applied context has been published and presented at a leading conference in energy management and consumption at MIT [17].

References

1. Bahner, J.E., Elepfandt, M.F., Manzey, D.: Misuse of diagnostic aids in process control: the effects of automation misses on complacency and automation bias. Proc. Hum. Factors Ergon. Soc. Ann. Meeting **52**, 1330–1334 (2008)
2. Balog, M., Gaunt, A.L., Brockschmidt, M., Nowozin, S., Tarlow, D.: DeepCoder: learning how to write programs (2017). Available via DIALOG, arXiv:1611.01989. Accessed 8 March 2017
3. Bickmore, T., Cassell, J.: Relational agents. In: Proceedings of the SIGCHI Conference on Human Factors in Computing Systems—CHI '01, New York (2001)
4. Botsman, R.: Who Can You Trust?: How Technology is Rewriting the Rules of Human Relationships, p. 183. Perseus Books, Cambridge (2017)
5. Bradshaw, J.M., Hoffman, R.R., Woods, D.D., Johnson, M.: The seven deadly myths of autonomous systems. IEEE Intell. Syst. **28**(3), 54–61 (2013)
6. Buckley, L., Kaye, S.A., Pradhan, A.K.: Psychosocial factors associated with intended use of automated vehicles: a simulated driving study. Accid. Anal. Prev. **115**, 202–208 (2018). https://doi.org/10.1016/j.aap.2018.03.021
7. Cassell, J., Bickmore, T.: External manifestations of trustworthiness in the interface. Commun. ACM **43**, 50–56 (2000)
8. Chien, S.-Y., Lewis, M., Semnani-Azad, Z., Sycara, K.: An empirical model of cultural factors on trust in automation. Proc. Hum. Factors Ergon. Soc. Ann. Meeting **58**, 859–863 (2014)
9. Chokshi, N.: Is Alexa listening? Amazon Echo sent out recording of couple's conversation (2018). Available via NYT, https://www.nytimes.com/2018/05/25/business/amazon-alexa-conversation-shared-echo.html. Accessed 25 May 2018
10. Dzindolet, M., Peterson, S., Pomranky, T., Beck, P.: The role of trust in automation reliance. Int. J. Hum. Comput. Stud. **58**, 697–718 (2003)
11. de Visser, E.J., Pak, R., Shaw, T.H.: From 'automation' to 'autonomy': the importance of trust repair in human-machine interaction. Ergonomics **0139**, 1–19 (2018)
12. Dijkstra, J.J., Liebrand, W.B.G., Timminga, E.: Persuasiveness of expert systems. Behav. Inf. Technol. **17**, 155–163 (1998)
13. Drnec, K., Marathe, A.R., Lukos, J.R., Metcalfe, J.S.: From trust in automation to decision neuroscience: Applying cognitive neuroscience methods to understand and improve interaction decisions involved in human automation interaction. Front. Hum. Neurosci. **10** (2016). https://doi.org/10.3389/fnhum.2016.00290
14. Endsley, M.R.: From here to autonomy: Lessons learned from human-automation research. Hum. Factors: J. Hum. Factors Ergon. Soc. **59**, 5–27 (2017)
15. Galdon, F., Wang, S.J.: From apology to compensation; a multi-level taxonomy of trust reparation for highly automated virtual assistants. In: Proceedings of the 1st International Conference on Human Interaction and Emerging Technologies (IHIET 2019) Conference, 22–24 August. Nice, France (2019)
16. Galdon, F., Wang, S.J.: Addressing accountability in highly autonomous virtual assistants. In: Proceedings of the 1st International Conference on Human Interaction and Emerging Technologies (IHIET 2019) Conference, 22–24 August. Nice, France (2019)

17. Galdon, F., Wang, S.J.: Optimising user engagement in highly automated virtual assistants to improve energy management and consumption. In: Proceedings of the 2019 Applied Energy Symposium AEAB Conference Proceedings, MIT, pp. 22–24 (2019)
18. Garcia, D., Kreutzer, C., Badillo-Urquiola, K., Mouloua, M.: Measuring trust of autonomous vehicles: a development and validation study. In: Stephanidis, C. (ed.) HCI International 2015 - Posters' Extended Abstracts, vol. 529, pp. 610–615 (2015)
19. Griffin, S.: Facebook's artificial intelligence robots shut down after they start talking to each other in their own language (2017). Available via INDEPENDENT, https://www.independent.co.uk/life-style/gadgets-and-tech/news/facebook-artificial-intelligence-ai-chatbot-new-language-research-openai-google-a7869706.html. Accessed 31 Jul 2017
20. Hancock, P.A.: Imposing limits on autonomous systems. Ergonomics **60**(2), 284–291 (2017)
21. Hancock, P.A., Billings, D.R., Schaefer, K.E., Chen, J.Y.C., de Visser, E.J., Parasuraman, R.: A meta-analysis of factors affecting trust in human-robot interaction. Hum. Fact.: J. Hum. Factors Ergon. Soc. **53**, 517–527 (2011)
22. Hoff, K.A., Bashir, M.: Trust in automation: integrating empirical evidence on factors that influence trust. Hum. Factors: J. Hum. Factors Ergon. Soc. **57**, 407–434 (2015)
23. Jian, J., Bisantz, A., Drury, C.: Foundations for an empirically determined scale of trust in automated systems. Int. J. Cognit. Ergon. **4**, 53–71 (2000)
24. Johnson, C.D., Miller, M.E., Rusnock, C.F., Jacques, D.R.: A framework for understanding automation in terms of levels of human control abstraction, pp. 1145–1150. In: IEEE International Conference on Systems. Man, and Cybernetics (SMC), pp. 5–8. Canada, Banff (2017)
25. Kaber, D.B.: Issues in human-automation interaction modeling: presumptive aspects of frameworks of types and levels of automation. J. Cognit. Eng. Decis. Mak. **12**(1), 7–24 (2018)
26. Kelly, C., Boardman, M., Goillau, P., Jeannot, E.: Guidelines for Trust in Future ATM Systems: A Literature Review; Reference No. 030317-01; EUROCONTROL: Brussels, Belgium (2003)
27. Kurzweil, R.: The Singularity Is Near: When Humans Transcend Biology. Penguin (2005)
28. Lee, J.D., Moray, N.: Trust, self-confidence, and operator's adaptation to automation. Int. J. Human-Comput. Studies **40**, 153–184 (1994)
29. Lee, J., Moray, N.: Trust, control strategies and allocation of function in human-machine systems. Ergonomics **35**, 1243–1270 (1992)
30. Lee, J.D., See, K.A.: Trust in automation: designing for appropriate reliance. Hum. Factors **46**, 50–80 (2004)
31. Leslie, D.: Understanding artificial intelligence ethics and safety: a guide for the responsible design and implementation of AI systems in the public sector. The Alan Turing Institute (2019). https://doi.org/10.5281/zenodo.3240529
32. Lewis, M., Sycara, K., Walker, P.: The role of trust in human-robot interaction. In: Abbass, H.A., Scholz, J., Reid, D.J. (eds.) Foundations of Trusted Autonomy, pp. 135–159. Springer Open, Australia (2018)
33. Madhavan, P., Wiegmann, D.A.: Similarities and differences between human-human and human-automation trust: an integrative review. Theory Issues Ergon. Sci. **8**, 277-301 (2007)
34. Madsen, M., Gregor, S.: Measuring Human-Computer Trust, 12 p. Central Queensland University, Gladstone (2000). Available via CITESEERX, https://cradpdf.drdc-rddc.gc.ca/PDFS/unc17/p520342.pdf. Accessed 15 Jan 2019
35. Manzey, D., Bahner, J.E., Hueper, A.-D.: Misuse of automated aids in process control: complacency, automation bias and possible training interventions. Proc. Hum. Factors Ergon. Soc. Annu. Meeting **50**, 220–224 (2006)
36. Mayer, R.C., Davis, J.H.: The effect of the performance appraisal system on trust for management: a field quasi-experiment. J. Appl. Psychol. **84**, 123–136 (1999)
37. Moray, N., Inagaki, T.: Laboratory studies of trust between humans and machines in automated systems. Trans. Inst. Meas. Control **21**, 203–211 (1999)
38. Moray, N., Inagaki, T., Itoh, M.: Adaptive automation, trust and self- confidence in fault management of time-critical tasks. J. Exp. Psychol. Appl. **6**, 44–58 (2000)

39. Mosier, K.L., Skitka, L.J.: Human decision makers and automated decision aids: made for each other. Autom. Hum. Perform. Theory Appl. 201–220 (1996)
40. Muir, B.M.: Trust between humans and machines, and the design of decision aids. Int. J. Man Mach. Stud. **27**, 527–539 (1987)
41. Muir, B.M.: Trust in automation: Part I. Theoretical issues in the study of trust and human intervention in automated systems. Ergonomics **37**, 1905–1922 (1994)
42. Nass, C., Steuer, J., Tauber, E.R.: Computers are social actors. In: Proceedings of the SIGCHI conference on human factors in computing systems, pp. 72–78, New York (1994)
43. Parasuraman, R., Manzey, D.: Complacency and bias in human use of automation: an attentional integration. Hum. Factors **52**, 381–410 (2010)
44. Parasuraman, R., Riley, V.: Humans and automation; use misuse, disuse, abuse. Hum. Factors **39**, 230–253 (1997)
45. Parasuraman, R., Sheridan, T.B., Wickens, C.D.: Situation awareness, mental workload and trust in automation: viable, empirically supported cognitive engineering constructs. J. Cogn. Eng. Decis. Mak. **2**, 140–160 (2008)
46. Parasuraman, R., Sheridan, T.B., Wickens, C.D.: A model for types and levels of human interaction with automation. IEEE Trans. Syst. Man Cybern. A Syst. Hum. **30**, 286–297 (2000)
47. Reeves, B., Nass, C.: The Media Equation: How People Treat Computers, Television, and New Media Like Real People and Places. Cambridge University Press, New York (1996)
48. Riley, V.: A general model of mixed-initiative human-machine systems. Proc. Hum. Factors Soc. Ann. Meet. **33**, 124–128 (1989)
49. Sarter, N.B., Woods, D.D.: Team play with a powerful and independent agent: operational experiences and automation surprises on the Airbus A-320. Hum. Factors **39**(4), 553–569 (1997)
50. Schaefer, K.E., Billings, D.R., Szalma, J.L., Adams, J.K., Sanders, T.L., Chen, J.Y., Hancock, P.A.: A Meta-Analysis of Factors Influencing the Development of Trust in Automation: Implications for Human-Robot Interaction. Defense Technical Information Center, Fort Belvoir (2014). https://doi.org/10.21236/ADA607926
51. Sheridan, T.B., Verplank, W.L.: Human and Computer Control of Undersea Teleoperators. Defense Technical Information Center, Fort Belvoir (1978). https://doi.org/10.21236/ADA057655
52. Simpson, A., Brander, G.N., Portsdown, D.R.A.: Seaworthy trust: confidence in automated data fusion. In: Taylor, R.M., Reising, J. (eds.) The Human-Electronic Crew: Can we Trust the Team, pp. 77–81. Defence Research Academy, Hampshire (1995). Available via DTIC, http://www.dtic.mil/dtic/tr/fulltext/u2/a308589.pdf. Accessed 30 Sep 1994
53. Sciuto, A., Saini, A., Forlizzi, J., Hong, J.I.: "Hey Alexa, What's Up?": a mixed- methods studies of in-home conversational agent usage. In: Proceedings of the 2018 on Designing Interactive Systems Conference 2018 - DIS '18, pp. 857–868 (2018)
54. Trimble, T.E., Bishop, R., Morgan, J.F., Blanco, M.: Human factors evaluation of level 2 and level 3 automated driving concepts: past research, state of automation technology, and emerging system concepts. (Report No. DOT HS 812 043). National Highway Traffic Safety Administration, Washington, DC (2014)
55. Wang, L., Jamieson, G.A., Hollands, J.G.: Trust and reliance on an automated combat identification system. Hum. Factors **51**, 281–291 (2009)
56. Westin, C., Hilburn, B., Borst, C.: Mismatches between automation and human strategies: an investigation into future air traffic management (ATM) decision aiding. In: Proceedings of the International Symposium on Aviation Psychology (ISAP) 2013 Conference, Wright State University, Dayton, Ohio, USA (2019)
57. Wickens, C.D., Dixon, S.R.: The benefits of imperfect diagnostic automation: a synthesis of the literature. Theor. Issues Ergon. Sci. **8**, 201–212 (2007)
58. Wickens, C.D., Li, H., Santamaria, A., Sebok, A., Sarter, N.B.: Stages and levels of automation: an integrated meta-analysis. In: Proceedings of the Human Factors and Ergonomics Society 54th Annual Meeting, pp. 389–393 (2010)
59. Yagoda, R.E., Gillan, D.J.: You want me to trust a ROBOT? The development of a human-robot interaction trust scale. Int. J. Soc. Robot. **4**, 235–248 (2012)

Chapter 15
Introducing the Concept of "Digital-Agent Signatures": How SSI Can Be Expanded for the Needs of Industry 4.0.

Alexander Pfeiffer and Mark Bugeja

Abstract Qualified digital/electronic signatures from human employees or legal entities can be used to generate sub-signatures for AI-powered software or robots/bots at the moment the dedicated tasks are being executed. This makes it not only possible to determine who gave the order, but also to store milestones of the respective order on a decentralised digital ledger (Blockchain). Smart Contracts can map further relevant functions such as the processing of contract-relevant data (e.g. logistics or insurance-related data), especially if, for example, different departments of one company or more companies are involved in the process. Furthermore, modern encryption technologies ensure that only the information can be read by those persons or companies for whom it is intended. Approaches such as "pruning" can offer the possibility to store private information only up to certain block height, after which only the proof of the transaction per se remains. This procedure could help Industry 4.0 processes not only being conducted efficiently but above all within a comprehensible legal framework. It can go even one step further, through digital-agents with appropriate classification clarifying the authorisation levels of additional AI-agents and generating such sub(sub)-signatures, on behalf of their original principals, to initiate further (needed) operating steps, which are recorded with the aid of specific Blockchain technologies.

A. Pfeiffer (✉)
Massachusetts Institute of Technology (MIT), Cambridge, MA, USA

Faculty of Education, Arts and Architecture, Danube-University Krems (DUK), Krems an der Donau, Austria
e-mail: pfeiffer@alexpfeiffer.at

A. Pfeiffer · M. Bugeja
Faculty of ICT, Department of AI, University of Malta, Msida, Malta
e-mail: mark.bugeja@um.edu.mt

A. Dingli et al. (eds.), *Artificial Intelligence in Industry 4.0*,
Studies in Computational Intelligence 928,
https://doi.org/10.1007/978-3-030-61045-6_15

15.1 Introduction

An 11-year-old purchases a new high-class gaming PC via Alexa voice command; An unintentional reservation in a luxury restaurant, made by a google-assistant; or 1000 Euro overcharged taxes, due to an error in AI-assisted accounting software. These are all cases, which are solved nowadays due to well-written terms and conditions, a helpful service hotline or in the worst case in court.

Now imagine these kinds of problems on a larger scale. An unauthorized person orders production machines via an AI-assisted purchasing software; unplanned business trips, booked by a digital assistant without knowing who gave the order initially; a minor software error that miscalculates the tax by "a 0", or production machines that have not been operated per their intended use. Most of these cases are unlikely to be solved on a goodwill basis but will have more significant consequences. Often, however, one will have to ask first who should be held responsible. The ever-faster technological progress is not only a blessing, e.g. through cost savings due to efficient production, but also brings problems with it. Issues that are not solved at a social, political, technological or legal level.

To put the problem into perspective: We currently lack a trustworthy solution to show orders given to "AI-assisted Software Solutions", "AI-driven language assistants", to AI-assisted robots/bots or even androids" between all parties involved. The problem becomes more prominent, the more partners are involved and whether they go beyond companies, municipalities, countries or even continents. Trust is described by Jósang [48] as a subjective belief in the reliability, honesty and security of an entity on which we depend for our welfare, and these entities contain software, hardware, data, people and organizations.

A possible solution to improve the situation sustainably is to combine the following technologies, to show who gave which order to the AI-Systems described above and point out the current status of the order fulfilment.

1. Qualified digital signatures or other accepted and trusted identity verification solution.
2. Blockchain technologies as a (partially) decentralized location for storing relevant data for the particular order.
3. Smart contracts to map more complex processes between different work steps and/or involved partners.

In the following section, various AI-systems are summarised under the term AI-assisted digital-agents, or simply digital-agents. This concept describes any AI system that performs activities on the order of a person, legal entity or another AI system with procurement rights.

The following section delves into a history of digital agents.

15.2 Digital Agents

Wooldridge [45] points out that there is no generally agreed definition of agents. He proposes the following definition, which reflects a revision of his thoughts from 1995 [43]:

> An agent is a computer system that is situated in some environment, and that is capable of autonomous action in this environment in order to meet its design objectives.

But he also says that this definition does not yet reflect the degree of autonomy of an agent nor the space in which it is located.

Monostori et al. [42] describes the role of agents along with other factors as follows. Digital agents should:

- Have a purpose to fulfill,
- Perform autonomous behavior and control both of their actions within the environment,
- Perform real-time information processing and adapt themselves to new situations,
- Prioritize events in accordance with their preferences,
- Exhibit intelligence, to some degree, from applying fixed rules to reasoning, planning, and learning capabilities,
- Interact with their environment in which they are operating, including the interaction with other agents,
- Be adaptive, that is, capable of tailoring their behavior to the changes of the environment without the intervention of their designer,
- Work as genuinely and transparently as possible, and
- Be credible and trustworthy in providing information to others.

Clearly spaces can exist where more than one agent exists. Huhns and Stephens [44] describe the conditions for such environments:

- Multiagent environments provide an infrastructure specifying communication and interaction protocols.
- Multiagent environments are typically open and have no centralized designer.
- Multiagent environments contain agents that are autonomous and distributed, and may be self-interested or cooperative.

Burden and Savin-Baden [49] define four different types of "AI-Systems", which can be adopted to the concept of a digital agent:

- Simple Algorithms—probably 99% of most computer programs, even complex Enterprise Resource Planning (ERP) and Customer Relationship Management (CRM) systems since they are highly linear and predictable.
- Complex Algorithms—programs such as, but not limited to, machine learning, deep learning, neural networks, Bayesian networks and fuzzy logic where the complexity of the inner code starts to move beyond simple linear relationships. Many systems currently referred to as AI sit here.

- Artificial General Intelligence—closer to what the public image of AI is, a system that can be applied to a wide range of problems and solve them to a similar level as a human.
- Artificial Sentience—beloved of science-fiction, code which ‘thinks’ and is ‘self-aware’.

Digital agents can be used for many purposes, but they have a significant impact on Industry 4.0.

15.3 Industry 4.0

Oztemel [40] defines Industry 4.0 as the transformation from machine dominant manufacturing systems to digital ones. He notes that this transformation will be opening the door to implanted technologies to the human body, wearable internet, cooperating and coordinating machines, self-decision-making systems, autonomy problem solvers, learning machines. He also notes that machines (which are subsumed as digital-agents by the authors) will even start to play the role of a decision board member having all the rights to make the decisions. Wang et al. [39] on the other hand, describes the main drivers of Industry 4.0 as mainly AI-Systems hand in hand with Internet of Things (IoT), Augmented reality (AR), additive manufacturing and cloud computing. With these technologies, Wang says, manufacturing systems can monitor physical processes and make real-time smart decisions. These developments are also shown in the literature review by Esmaeilian et al. [41].

An important factor is the authentication (identity) of digital agents, especially for Industry 4.0 use-cases. The authors suggest to combine digital-agents with qualified digital/electronic signatures (or other trusted digital identity solutions that use sophisticated technology like blockchain and therefore enable smart contracts). The authors refer to this concept in the following as a digital-agent signature. However, in the next section an overview on the forms that the digital identity can take is presented.

15.4 Digital Identity

Digital or electronic identity is an essential component: with this identity the digital-agents are entrusted with tasks from a certain person or in the name of a certain company and the quality of this identity will decide whether, for example, a digital agent can interact with other digital agents or persons outside the company.

This identity can come from various sources. An employee may have been assigned his digital identity by his employer, but he can also use his identity obtained through a service provided by the government. This is because a citizen card identifies the person whilst a database stored centrally at the respective company, provides the user rights and activates them after login with the digital/electronic signature. The signature can also have been assigned by an external company that specializes in the creation of these signatures or it can also be generated through an interface like Facebook Connect.

All these different sources offer a range of levels of trust, inside the institution where the signature is principally used and especially when interacting with other parties. This level of trust or its valuation is ultimately determining how far the authorization of the respective digital/electronic signature goes.

This means that first, it is necessary to discuss digital identity, how it is expressed, and when it becomes necessary:

- Human with computer interaction: The average person logs in 7–25 times per day [46]. In the simplest form a "user" plus "password" system is being used. This can either be assigned by a system administrator (human or software) or by the user himself. The username can be selected at login (this would be the most insecure option), or it must also be entered (and known) by the user, which would be the slightly more secure variant. Pidel and Neuhaus [47] describe four different types of passwords Pass phrases: multiple words separated by spaces ("cats are fantastic friends");—Typical passwords: a single (in our case, English) word with letters replaced by numbers and special characters ("C@terp!11ar2018"); Randomized: a fully randomized string of characters, such as those used by password managers ("nBqzEcP2A}Q8,jG") and—Bad passwords: passwords often seen in password leaks ("password123").
  Modern log-in systems also use additional authentication. For example, via "Two Factor Authentication", which means you need a second device or account to which a code word is sent (text-message or TAN) or a code is generated every 30 s (a well-known system has been developed by google, the so called google authenticator[1]). Another possibility would be to use biometric data such as fingerprint or face recognition.[2] On the server side, the password should not be stored by the software operators in text form but encrypted in such a way that the original password is never visible, and only the process of setting a new password can be initialised. The latter seems obvious, but cases involving Twitter,[3] Facebook [4] or T-Mobile [5] have shown that this major security hole still exists.
- Computer to computer interaction: (digital-agent with digital-agent, digital-agent with software, software with digital-agent). In this case, too, a digital identity in

[1]Google Authenticator: https://www.google.com/landing/2step/ (19.12.2019).

[2]https://www.Intego.com—from Touch-ID to Face-ID: https://www.intego.com/mac-security-blog/apple-security-touch-id-vs-face-id/ (19.12.2019).

[3]Wired on Facebook storing PW in plain https://www.wired.com/story/facebook-passwords-plaintext-change-yours/ (19.12.2019).

[4]Hackread.com on the twitter Password bug: https://www.hackread.com/bug-stored-twitter-passwords-in-plain-text-change-password/ (19.12.2019).

[5]Reddit about T-Mobile Austria storing the first 4 digits of Passwords in plain-text: https://www.reddit.com/r/sysadmin/comments/8aem4n/tmobile_plaintext_password-data_breach_thought_to/ (19.12.2019).

the sense of a proof of entitlement must be provided, in the best case scenario this process can track right back to the original source. It is important, however, that the instruction to the software or digital-agent is guaranteed by the most secure authentication possible. On the other hand, this is not the case, if a simple user password system has been used following self-authentication.
- Computer to human interaction: This takes place when a digital-agent approaches a human to enter further data, to perform a production step or to mark a work as completed. Like the previous case, it is also important that it is verifiable from which source, or from which sources, the initial order originated.

Ronchi (2019) describes in his book e-citizen which types of passwords exists and how proof of identity is typically delivered, e.g. by additional data like birthday, grandmothers name and other such cases. However most of this information is self-provided by the user himself beforehand to the system or system-administrator.

Finally, there is one point that needs to be considered and is overlooked by many—if the necessary trust goes beyond activities in one's own institution, or with processes that do not only affect one personally, then it is extremely important to ask the question who, with which certainty and by using which resources confirms and establishes the original identity in the first place?

This is a question of trust, which qualified digital/electronic signatures, issued by the national governments, if accepted across national borders, might solve.

15.5 Qualified Digital/Electronic Signatures, Based on the Example of eIDAS EU Regulation Compliant Approaches (e.g. From the Austrian A-Trust GmbH, LuxTrust or E-ID Malta)

Within the European Union, different countries have worked on various solutions for digital/electronic qualified signatures. However, the different approaches share the same basic principles. On the one hand, these approaches are based on multi-factor authentication. Meaning they use at least two of the following elements:

1. knowledge (something only the user knows),
2. possession (something only the user has),
3. inherence (something the user is).

On the other hand, these signatures have been issued by "Trust Centers", which have received authorization from the respective government.

Digital/electronic signature are mostly used for the following purposes, as e.g. described on the webportal of the LuxTrust, the Trust Center of Luxemburg.[6]

[6]Luxtrust.com about electronic identity: https://www.luxtrust.com/trust-services/electronic-identity/.

1. authenticate yourself on a web-portal (company, banking, insurance, governments)
2. sign electronically a document that is legally binding
3. seal electronically data
4. exchange data with third parties.

One difference between the various approaches is the medium on which the certificates are retrieved. For example, Luxembourg used dongles/tokens but recently started their LuxTrust Mobile Service, moving their services from dongles/tokes to a smartphone application, Malta uses a smart card[7] and Austria currently has a dual solution where citizens and institutions can use either a smart card with a dedicated reader or their mobile phone. The latter is known on the market as a "Handy-Signatur", whereby "Handy" is the colloquial word for mobile phone, or now smartphone.[8] The solution on smart cards will be discontinued for citizens in Austria, as it can no longer be activated on the e-card (the social security card in Austria). This is due to the great success of the "Handy-Signatur". For companies or professional groups such as lawyers or notaries, however, the smart card solution remains a reasonable option. Here, the actual possession of a smart card still makes sense and it can also be individually branded.

At this point, the question arises which persons or institutions are authorized to issue digital/electronic signatures to private individuals, public bodies or companies? Trust Centers usually work together with reputable partners. These are partners who are considered trustworthy from the point of view of the population and the industry. These can be banks, insurance companies, notaries, public authorities or Internet services directly related to ministries, such as those portals where citizens report and pay their taxes. Persons who are authorized to issue signatures must undergo training and/or examination procedures. In Austria, for example, this is the training to become a Registration Officer.[9]

However, in Austria, a major change is currently taking place, with E-ID being introduced in autumn 2020. It may only be issued by the passport authorities and police stations. The E-ID is considered a further development of the citizen card and "Handy-Signatur". Previous certificates will remain valid until they reach their respective expiry date. The change makes the digital/electronic digital signature one step more trustworthy due to the extremely restrictive authorization of who may issue them. On the other hand, this could of course lead to bottlenecks in issuance and delays in distribution within the Austrian population.

From a developer's perspective, the website on the Austrian E-ID already offers instructions for early adopters and the possibility of having the connection of the E-ID to the respective developed application checked [1]. A check is a prerequisite for a successful connection and adaptation. This leads to a strong increase in security and trust for future users.

---

[7] Identity Malta about services using E-ID in Malta: https://identitymalta.com/services/.

[8] A-Trust.at about the "handy signatur": https://www.a-trust.at/de/Handy-Signatur/e-id/?b.

[9] https://wiki.a-trust.at/wiki/Default.aspx?site=Registration%20Officer.

Nevertheless, one of the biggest predicaments is that these are solutions that have all been developed separately in different countries. Therefore, until a few years ago, there was no possibility of recognising signatures of another Trust-Center (or EU country) beyond the national borders. The eIDAS Regulation is intended to remedy this situation and enable homogeneous recognition within the European Union or its Trust Center. It is important to note that this is a lengthy process and that recognition does not yet extend beyond that within the EU. It is therefore clear that a global solution is still a long way off [3–5].

A first step in the right direction was the European Signature Dialogue in 2017 [6]. A Trust invited the decision makers of the other European and non-European Trust Centers and signature services to coordinate the further common strategy.

In addition to identity service providers officially authorized by governments and homogenized by the eIDAS regulation, at least in the medium to long term, there is of course also a market for corporate service providers offering identity verification services; the Self Sovereign Identity verification movement using blockchain-technologies; or open source applications for registration and issuing of certificates by institutions such as universities. Besides, social media applications like Twitter or Facebook also like to play a role on the identity-verification market.

15.6 Social Media; Companies and Universities as Identity Providers/For ID Verification Services

Social media providers such as Facebook or Twitter offer private individuals, companies, journalists, politicians or artists the opportunity to publish content. It is also possible to run accounts for fictional characters, for example the company Marvel can run an FB account for the comic character Iron Man [7]. For this purpose, a profile is first created with the identity with which the content is to be published.

However, it is also possible to register under a false name. This is for example the case if someone wishes to protect his own identity, but also to cheat consciously, to deceive and offer false information (fake-news [8]) or to get money for clicks (click-farm operators [9]).

Nevertheless, identity plays a big role via social media. The rules for handling identities are laid down in the terms and conditions or in the respective national legislation. Facebook introduced "Facebook Connect" [10] in 2008. This enables other platforms that have user accounts in addition to their own database, or even without their own user database, to offer individual services and outsource login management. The advantages are mostly that users are generated faster for a service or the liability in case of data theft etc. is minimized. The disadvantage is the dependence on an external provider for the user data and ultimately the value of the user data for the calculation of a company value.

Twitter has introduced verified accounts [11], this service (currently on hold) allows persons or companies of public interest to get a blue tick after detailed verification by twitter. However, Twitter has often come under pressure, for example when the right-wing extremist Kessler got a blue tick [12]. Another way of verifying that a person is "real" is to write a private message or even a public tweet from the user's account on Twitter with information such as the code of the site for which they are signing up. This is often used in "grassroots" movements to determine identity. Conclusion: As long as the verification only works on the basis of one's own data or at most with a telephone number (even if in some countries an identity has to be shown even for pre-paid cards), ID statements via social media are to be classified as rather untrustworthy.

Universities and colleges either use simple login systems that are not considered very secure, commercial or governmental systems described later in this document, or implement their own systems to manage users, user groups, rights within the university and rights to third-party software or services. One such system is MIT Touchstone [13]. Every student, employee or frequent service provider receives its own digital signature, with which login options are defined, PDFs and contracts can be signed within the system, personal data can be changed, and even third-party providers and their services (e.g. students may use XFINITY free of charge [14]) are regulated, as long as the third-party providers have implemented Touchstone compliant login solutions. With Touchstone, MIT is securing its own infrastructure, services, contracts with third parties, and last but not least for students the registration for courses, access to grades, etc.

Digital/electronic identity is also notably important in the Industry 4.0 market. Especially in the face of constant progress and the current digital transformation.

Therefore, Know Your Customer (KYC) and Anti Money Laundering(AML) regulations through various industries have strongly stimulated identification checks in the digital world by private companies. Industries that now use such identity verification services include the financial services industry, ecommerce, travel-agents, telecommunication providers, dating platforms, and the igaming industry (TechUK 2019).

Identity checks usually work by having the user provide data such as a copy of an ID card, passport or driver's license plus a photo of themselves. This information is then compared with publicly available data and an identity score is generated [15, 16].

Some providers also offer different levels of verification. For example, if a user wants a higher limit on their credit card or stock or cryptocurrency exchange, they will need to make a video call with a company agent (using proprietary software provided by the identity verification agency). For example, in a multi-step procedure step 1 is the user's self-input, step 2 is the upload of the ID documents and step 3 is the video call. A possible step 4 allows further background checks on the person, for example, checking bank account balances. The background check of identity by banks can also lead to the bank itself acting as an identity provider, as the example of S-Identity of Erste Bank and Sparkassen in Austria shows [17]. Returning to the point of the good faith in a digital identity, the question of who verified it, who issued it, and how certain one can be that the identity is nearly 100% valid, it must now be decided whether employees of commercial companies who perform a simple verification based on scanned data or, in the best case, a video phone call, achieve the same level of verification as a signature created from a personal interview with an official.

For some applications this type of check will be sufficient and it is in any case necessary for applications for citizens of countries that do not yet have a digital identity issued by their government. There might also be countries whose government is not considered trustworthy, and therefore a verification from a private company is considered more suitable. One must always consider here: Where is the data stored? Who has access to the data? Can transactions be viewed, and if so, by whom? Furthermore, above all: how tamper-proof is the data stored, in particular—can an employee of the Corporate Service provider change a digital/electronic identity with a simple database entry? Moreover, if so, how is this change logged, and how difficult is it to override this log? Mahajan, Sperling and White (2019) describe in their article "Digital ID: The opportunities and risks" digital/electronic signatures as "dual use" technology, such as GPS or Social Media [18]. This is especially so when talking about frauds with digital/electronic ID probably made possible due to security vulnerabilities in the Corporate Service providers.

Responding to all of these questions from above a new concept of identity management has come up: Self Sovereign Identity (SSI), a concept that has recently surfaced in the blockchain community.

15.7 Self-sovereign Identity and Multi-source Identity

The Sovrin Alliance (2017, 2019 [19]), a foundation with the aim to promote the idea of Self-Sovereign Identity describes the concept as follows:

"Self-Sovereign identity (SSI) is a term used to describe the digital movement that recognizes an individual should own and control their identity without the intervening administrative authorities. SSI allows people to interact in the digital world with the same freedom and capacity for trust as they do in the offline world [...]

[...] Self-sovereign identity (SSI) brings the same freedoms and personal autonomy to the internet in a safe and trustworthy system of identity management. SSI means the individual (or organization) manages the elements that make up their identity and controls access to those credentials—digitally. With SSI, the power to control personal data resides with the individual, and not an administrative third party granting or tracking access to these credentials. The SSI identity system gives the stakeholder the ability to use their digital wallet and authenticate their own identity using the credentials they have been issued. They no longer have to give up control of personal information to dozens of databases each time hyou want to access new goods and services, with the risk of their identity being stolen by hackers. This is called "self-sovereign" identity because each person is now in control of their own identity—they are their own sovereign nation. People can control their own information and relationships. A person's digital existence is now independent of any organization: no-one can take their identity away."

Following Windley (2018 [20]) SSI is the next step following Multi-Source Identity (MSI). Windley describes the concept of MSI as follows:

"MSI allows multiple credentials from multiple providers to be brought to bear, flexibly and conveniently, in a situation where trusted attestations are needed for the participants in a workflow to make progress. In MSI, there are three players: credential issuers, credential holders, and credential verifiers. Any person or organization can play any or all of the roles." Windley says not every identity is instantly true. It depends on the various identity verifiers who assign trust levels to identities. For Windley evidence of identity is key, and how much evidence is needed depends on the usage of the identity and the risk involved. The transformation from MSI to SSI takes take effect when:

"Multi-source identity becomes self-sovereign identity (SSI) when the individual is able to control the credentials and use them in a privacy-preserving manner whenever and where ever they want. Privacy is a critical feature of SSI because without privacy, there is no control. In SSI, the identity owner must be able to control who sees what and that means that privacy is a fundamental property of the architecture for SSI."

Self-sovereign identity therefore means: The individual or organization controls and manages their identity.

Christopher Allen [21] describes the ten key principles for a successful SSI approach. The ten principles are divided into three main areas: Security, Controllability and Portability:

Security	**Controllability**	**Portability**
Protection	Existence	Interoperability
Persistence	Persistence	Transparency
Minimisation	Control	Access
	Consent	

Baars (2016) describes in his MA-Thesis (University of Twente) "Towards Self-Sovereign Identity using Blockchain Technology" the important role of Blockchain Technology for SSI. He has designed a conceptual prototype to address the KYC problem at banks. His conclusion is as follows:

"The design is dependent on a storage layer where everyone can be sure that if claims are made, it is non-reputable that the claim has been made and by what entity. In our solution, blockchain technology provides this functionality by replacing the need for a central authority using cryptographically secured consensus algorithms. To become useful, all relevant actors need to adopt the system. Our case study showed that proprietary solutions are undesirable if trust is important"

In Austria the Österreichische Staatsdruckerei GmbH"(Austrian State printing house, formerly owned by the government, privatized since 1999) [22], which is producing high standard security solutions, e.g. the biometric chips used in travel documents , is currently working on My Identity App (MIA) [23]. MIA offers a highly secure electronic identity (eID) to users by combining traditional ID documents with a digital identity in a smartphone app that operates on any platform. With MIA, users can transparently identify and authenticate themselves in both the physical and digital world. The Österreichische Staatsdruckerei claims that user can rest assured that security, privacy, data protection, usability and trust placed in the app have been considered in equal parts.

The next development step is already underway. The "Austrian Log-in Alliance" was founded in April 2019 [24]. It includes leading banks, the Staatsdruckerei GmbH, other technology companies, the trade association and notaries. Their product is called "myIDsafe". The idea behind: An independent offer for a personal digital/electronic identity is to be created for private individuals as well as for the economy and industry 4.0, secured by the encryption mechanisms of blockchain technologies. The human being as a sovereign owner of his or her own digital identity will be placed at the centre of attention. In this sense the "Self-Sovereign Identity" the users personal data is completely under his/her control. There is no central instance that manages or controls this data. The users can manage everything from their smartphone, while ensuring that no private data itself is stored on the mobile device. In this way, the log-in alliance partners want to create an Austrian, cross-sector and at the same time open trust network, which at the same time eliminates competitive disadvantages compared to US networks and provides a secure infrastructure for digital business processes. However, the problem might be the international acceptance of an Austrian solution, which in fact sees itself as competing with overseas providers.

In most of the concepts mentioned above, only the wording "the blockchain" is used without going into detail. However, blockchain as a trust-building link is without doubt a key element of SSI. In the next section, a closer look at blockchain technologies is presented.

15.8 Blockchain Technologies and Smart Contracts

Immutability to changes made retroactively (generally speaking and ignoring software bugs or hacks that have already affected some systems [25–27]) make Blockchain systems the perfect technology to secure date (Atzei et al. 2017). Keeping data trustworthy, secure, and manipulation-proof has become an increasing issue in e-commerce and industry applications, due to the already discussed ongoing digitalisation and now introduction of AI-systems.

Due to their ability to store data in a (at least in part) decentralized, transparent, and manipulation-proof way, Blockchain-based technologies can provide solutions to this problem. While the allocation of digital/electronic identities and how they can be used has been discussed in detail up to now, we will now discuss how the data processes that were started with them can be stored securely (Schmidt 2019).

In the early 1980s, Lamport et al. (1980, 1982) described the Byzantine generals' problem and outlined possible solutions in their papers 'Reaching Agreement in the Presence of Faults' and 'The Byzantine Generals Problem'.

The name refers to the attack on Constantinople in 1453, during which the city had to be attacked simultaneously from several sides due to its strong fortifications.

The generals who were to coordinate the attack communicated with each other by messenger, as they could not be in the same place at the same time. However, some of these generals were prone to send false and contradictory messages to other generals on purpose, as they wanted to gain a personal advantage by discrediting their fellow combatants in the eyes of the sultan. In effect, while there was a constant flow of messages in all directions, the receivers could never be sure whether they could trust the message and the person who had it delivered to them. Furthermore, even if one could have been sure that the signature of the sender is correct, how can one be sure that no manipulation of the data was made on the way between two generals? This is exactly the problem which is now described in this chapter and why blockchain technologies are an important element of the concept described.

Grech and Camilleri (2017) describe the positive effects of Blockchain Technologies as follows:

"From a social perspective, Blockchain technology offers significant possibilities beyond those currently available. In particular, moving records to the Blockchain can allow for:

- Self-sovereignty, i.e. for users to identify themselves while at the same time maintaining control over the storage and management of their personal data;
- Trust, i.e. for a technical infrastructure that gives people enough confidence in its operations to carry through with transactions such as payments or the issue of certificates;
- Transparency and provenance, i.e. for users to conduct transactions in knowledge that each party has the capacity to enter into that transaction;
- Immutability, i.e. for records to be written and stored permanently, without the possibility of modification;
- Disintermediation, i.e. the removal of the need for a central controlling authority to manage transactions or keep records;
- Collaboration, i.e. the ability of parties to transact directly with each other without the need for mediating third parties."

The concept of Blockchain, as we know it today, derives from Satashi Nakomoto's Whitepaper "Bitcoin: A Peer-to-Peer Electronic Cash System", published in late 2008. (Satashi Nokamoto is a pseudonym, it is not known to the general public who is behind this name.) Originally intended to create a non-manipulable account book to represent the possession of digital tokens, which in turn are traded for money on an exchange, it is now about the technology behind it and what applications can possibly be developed with it. The idea of using the Bitcoin Blockchain for more than "proof of payment transactions" arose from the fact that you can attach text messages to a transaction. To create an account book of any imaginable transaction, a fraction of Bitcoin was sent to an address (so-called Satoshis) and the text to be recorded was attached to it as a text message and thus stored forever on Blockchain. This basic principle is used, for example, by Blockcerts to store learning credentials [28]. Since one is, however, strongly limited in the range of possible applications, if information of all applications is stored only as a text message by the same token. Moreover, since this token was not originally intended for other applications outside the payment flow, the idea was to develop a network in which sub-tokens (metatokens) can be generated for the respective specific application. The blockchain systems nxt and Ethereum are particularly noteworthy in this context from a historical as well as actual perspective. Next [29] because it was the first Proof of Stake (PoS) system, a different approach in the field of consensus finding to Proof of Work (PoW), which is considered very environmentally friendly. Ethereum [30] because it is the second most important blockchain system after Bitcoin.

However, both blockchain-systems mentioned also have another special feature. The possibility to develop smart contracts. While Ethereum makes this possible with Solidity, a specially created programming language, nxt uses so-called lightweight smart contracts in Java. This offers less design freedom, but a higher level of security. In principle, smart contracts is nothing else than the digital, (partially) automated and secure processing of contract scenarios.

It is also very important to mention that generally blockchains can be classified into three major categories (Wan et al. 2019):

1. Private Blockchain: is basically a closed system and exclusively operated within organizations, businesses, or government structures. No information is disclosed to the outside world, except, where appropriate, evidence of a transaction that has taken place.
2. Consortium Blockchain: serves related parties have a common goal, and related organizations can join the consortium blockchain on common agreements. Again, no information is disclosed to the outside world, except, where appropriate, evidence of a transaction that has taken place.
3. Public Blockchain: has no restrictions regarding joining or leaving the blockchain. All information is public, although it is possible to store some information in encrypted form.

Due to the decentralized development, the still short time (10 years) of the Blockchain idea itself and the interest of industry and society even more recently, there is still no satisfactory standardized "wording" or "definition" of the different types of tokens [31–35], especially an urgently needed differentiation of utility tokens according to their use cases. This classification could ensure that a different legal perspective applies to different cases. Therefore Pfeiffer recently proposed the following segmentation regarding blockchain-based assets at the IMH conference [36]. The authors have revised this proposal and categorize Blockchain Assets as follows:

1. Cryptocurrencies: Tokens with the purpose to serve as currency supplement and to transfer monetary values. Traded on dedicated exchanges or over the counter. [Like Bitcoin (BTC), Litecoin (LTC), Ripple (XRP)]
2. Stable-Coin: Tokens with the purpose to serve as currency supplement and to transfer monetary values, with a nearly fixed conversion rate to a FIAT reserve currency.[Like Tether (USDT), USD Coin (USDC), AEUR (AEUR)]
3. Tradeable network maintenance utility tokens: Tokens which serve as a reward to maintain the network. Traded on exchanges. May also have aspects of other

token categories, e.g. additionally serve as cryptocurrency. [Like Ethereum (ETH), Ethereum Classic (ETC), NXT (NXT), Ardor (Ardr), EOS (EOS)]
4. Non free-tradeable network maintenance utility tokens: Not traded on exchanges, but their distribution can determine the power within a consortium-based (or private) network. [tokens that are not traded on exchanges but distributed from a smart contract or centralised authority, mostly with a fixed number of issued tokens from the beginning, like a typical PoS approach; this could be also a token that represents a right to vote for something (and its voting power)]
5. Tradeable utility tokens: Tokens that have a specific purpose, e.g. to represent a digital or a real good. The value of this asset is determined by supply and demand on a token trading exchange. [Like Steem (Steem), Augur (REP), or 'historic' approaches like Amps (AMP), NEXIUM (NXC)]
6. Fixed-price (including peer2peer price-negotiable) tradeable utility tokens: Tokens that have a specific purpose, e.g. to represent a digital or a real good. The value of the token is determined by the issuer, A third party that has a contract with the original publisher that allows them to set the prices, as a result of negotiations, or for example at a classic auction (as a restricted bargaining room). [Any specific token that represents e.g. a digital art piece, or a real life item of value (indicating its ownership)]
7. A (tradeable or non-tradable) Token representing a share of ownership or a share of contribution to something, leading e.g. a certain rewards. [Like KuCoin Shares (KCS), Huobi Token (HT) or Binance Coin (BNC)]
8. Non-freely tradeable utility tokens: These tokens store data, such as certificates, grades, ownership of a piece; fine art prints (e.g. limited edition prints, each with a unique number), or a last will; they can be a unique (singleton) token per record or a message attached to a specific token when sending. A separate series of tokens is generated for each different use case. Each set has its asset ID on the respective Blockchain. (the name of the series does not have to be unique, only the asset ID). This means: The moment a message is added to one of the tokens (from a series) and this token is sent, the connection of the token with the message and the rule that the token cannot be forwarded without the knowledge of the original sender becomes a unique process, which is identified by the unique transaction ID. Messages can be attached unencrypted or encrypted. This data is usually linked to a person or a property and is not (or only under specific circumstances) tradable. It is also linked to a specific wallet (e.g. of the recipient). The Singleton/Unique Token form of this category is similar to the concept of non-fungible tokens (NFTs). [e.g. Tokens that are used to store valuable information either in Token properties or as message attached, like storing the grades in the Serious Game for learning and assessment Gallery-Defender]

The connection between wallets (holding tokens) with an identity (of a person, a company, a club, a political party etc.) is therefore absolutely essential. Pfeiffer presented his concept of E-ID wallets in his keynote at the "eEducation Praxistage" in early 2019 [37].

15.9 E-ID Wallet

An E-ID wallet, is a wallet for blockchain-tokens, which is linked to one or more digital signatures of the owner(s).

Valid signatures include government issued signatures, any signature from a self-sovereign identity app, a signature issued by an educational institution, signatures issued by an identity verification company or a connection to a social media account. There are different levels of trust in the digital signature to be considered.

Depending on the selected signature type, the proof of the signature transaction is stored and displayed differently. The signature hash value can be published, for example, on a protocol page of the respective trust center, on the blockchain used by the SSI app, in which case, a token (including the private data as encrypted message) is sent from the (signed E-ID wallet) of the SSI app provider to the newly signed E-ID wallet of the user. Alternatively, it could be a published tweet, with the wallet address and the signature value, if the own Twitter account has been chosen for identity binding. Again, different levels of trust establishment require different measures to store the signature proof.

The signature chain can be retrieved for each E-ID wallet. For example, if an E-ID wallet is dedicated to a department of a company, you will see that the main account of the company and the person responsible for the department have signed. However, who originally signed the main account of the company? and how did the main person responsible for the department get this status. Whether the private data is publicly accessible or encrypted is of course always subject to the person or institution and their needs. In other words, whether it is essential that everyone can see who the wallet is assigned to, or if only persons, company-partners, or other departments of a company who gain access to this information should know the ownership.

However, the special feature of the E-ID Wallet is that, in addition to digital identities, it can and should also hold blockchain-based tokens and can therefore be used for utility tokens linked to an identity on the one hand, but also for cryptocurrencies as a form of payment with proof of identity (to counteract money laundering and other similar problems) on the other.

Now it is a matter of connecting digital signatures and E-ID wallets with digital agents, so that their distribution of tasks and progress is stored with the highest possible security and allows the digital agents to interact with third parties.

It should also be noted that some blockchain-token wallets already offer the possibility to name the wallet publicly (for example original Ardor wallet [37]), but here it is a pure self authentication.

15.10 Introducing: The Digital-Agent Signature

The basic idea of the digital agent signature is to connect the (digital) identity of the user of the digital-agent with the digital agent itself. The user uses a signature that is available in his SSI app. This allows the user to select the appropriate signature for the different applications. When the digital-agent is instructed to carry out an administrative task, such as monitoring and paying tax returns, a government-issued signature will be used. If the digital-agent is instructed to search for and purchase the best possible car insurance, a signature issued by an identity verification company is used, a digital-agent is instructed to compile and enroll in the best schedule for study, that signature issued by the university is used, and if the digital-agent is acting on behalf of the user on a love mediation platform, an even lower level of verification may be sufficient.

If the signature is used in a work context, a custom SSI app that is provided and digitally signed by the respective company can be used. This SSI app has only the signatures verifying the identity of the user which are authorized and accepted by the respective company. The SSI app also shows the user's position and rights within the company.

The particular feature of the digital-agent signature is that a token transfer is triggered during the signature process. From the E-ID wallet of the person or company for whom this person works to the digital-agent's E-ID wallet.

This token is a "Non-freely tradeable utility token", which cannot be sent without the consent of the sender account. A reason to resend the token could be to "burn" it, i.e. send it to an address where no private key is available. This could be useful after a minor task is completed or for a major task after e.g. a seven year wait (if government regulations requires to store certain date for this time length). The token contains either/and in the account properties or as (encrypted) text message all necessary information (rights and duties) which the signature process triggers. Duties are the tasks to be performed by the digital agent, and whether and how any intermediate performance steps or the performance of the task is stored (ideally again on blockchain if the activities are handled between external partners). Rights are those operations in which the digital-agent identifies itself with its own signature to start a process. This signature is linked to a signature chain with all previous signatures for the respective task and is therefore a unique ID that represents the entire process. Rights also mean the actions that the digital-agent is to execute under a smart contract.

To sum up: The digital agent signature combines self sovereign identity (e.g.: digital qualified signatures) with verified blockchain wallets (E-ID wallets) and non-tradable utility tokens as carrier medium for data and authorization to operate.

This not only provides a complete record of interactions with and between digital-agents and their tasks, it also ensures that the user can see who has given the orders for the actions. All this information can be kept private, either in whole or in part, with the ability to assign shared-keys for access rights. Only the hash value that a transaction has taken place should be public or at least shared between the consortium.

This new form of interaction, or rather the protocol of interaction between users and the digital agents, or the digital agents among themselves, opens up a large number of use cases in the Industry 4.0 area, since it is precisely here that authenticity, automation and maximum security play an important role.

15.11 Possible Use Cases in Industry 4.0

The authors would now like to list (non-taxative) some possible applications for the digital-agent signature in industry 4.0:

- Supply chain management and logistics: The digital-agent signature enables the production and logistics chain to be securely monitored. The corresponding data is visible for all parties in the production, delivery and sales process and ultimately for the end customer. Each in the form and depth necessary for a transparent identification process.
- Monitoring of production facilities and processes: Within a production facility, the production processes and the production halls can be monitored and, for example, automated steps can be taken to ensure that production runs optimally and safely.
- Quality assurance: Quality control can be logged automatically and, if necessary, suggestions proposed by the digital agents to can be initiated.
- Change in production capacity: Proposals of the digital agents to change production capacities, e.g. to achieve a production target or to adapt to a market situation, can be accepted in a more goal-oriented and efficient way due to the digital agent signature, the exact assignment of rights, the assigned possibilities for actions triggered by the AI and its logging.
- Settlement of fines, penalties and Insurance in the Industry 4.0 Sector: The digital agent signature enables penalty payments, other types of fines and insurance settlements to be processed more fairly, transparently and above all more quickly.
- Administration: The digital agent signature enables administrative processes within companies, but above all also between companies, or companies and authorities, to be handled efficiently and cost-effectively.
- Customer Service: Digital Agents can be used to provide customer service for standardized requests.
- Secure Oversight: Due to GDPR and data protection laws video footages and video data on sight cannot be viewed by individuals unless proof of criminal activity is given. A digital agent can be used to view this information securely and ensure that only individuals that have the key (local enforcement) can view this data.

15.12 Concerns, Conclusion and Outlook

In order for the concept of digital-agent signatures to be introduced in the next few years, a few more steps are necessary:

- Decision makers need to have confidence in digital signatures and blockchain technology, or as a first step, an understanding of these technologies.
- All approaches have to be developed using an open-source approach, and using strong decentralized systems for storage. Monetization is done by individual solutions, consulting and support.

- Processing of different (all common) blockchain systems and possibilities for digital identity verification (in the SSI apps).
- Interface development between AIs (digital-agents) and current computer systems, so that the digital agent signature can be used.
- A modern AI, blockchain and digital identity legislation.
- And, of course, possible downsides such as cyber security, hacks, social hacks or other technical developments such as quantum computers must also be taken into account.
- Furthermore, first journalists and media experts and subsequently the general public must be made aware of the described forms of technology (especially AI, digital signatures and blockchain) and their possible applications. This step is necessary to reduce fears on the one hand, but also to gain the necessary caution and media competence on the other hand. All of this is recommended to make the step into the age of digital agents successful. Also the limitations of the used systems must be considered.

The next step for the authors is to develop a conceptual prototype, linking the upcoming Austrian E-ID solution with Blockchain Wallets from the Ardor-Testnet System. With this setting a first demo for the digital-agent signatures can be established and further applied research conducted. It is also important to have a look in the past, research which prototypes of identity-verification between agents existed in the pre-blockchain area, what can be learned from them, and if the digital-agent signature might be the concept that would solve issues that occured.

References

1. Developers guidline to implement E-ID in future applications: https://eid.egiz.gv.at/secure/?page_id=259#signatur
2. The Austrian E-ID approach: https://eid.egiz.gv.at/secure/
3. MITA on eIDAS Regulation in Malta: https://www.mita.gov.mt/en/ict-features/Pages/2017/MITA-and-Identity-Malta-compliant-with-the-ET//SI-standard-on-electronic-signatures-and-infrastructures.aspx
4. Cryptomathic.com explaining the eIDAS regulation: https://www.cryptomathic.com/news-events/blog/qualified-electronic-signatures-for-eidas
5. Regulation (EU) No 910/2014 of the European Parliament and of the Council of 23 July 2014 on electronic identification and trust services for electronic transactions in the internal market and repealing Directive 1999/93/EC: https://eur-lex.europa.eu/legal-content/EN/TXT/?uri=uriserv:OJ.L_.2014.257.01.0073.01.ENG
6. A-Trust about the European signature dialogue: https://www.a-trust.at/de/presse/news/eidas-Handy-Signatur-ueber-oesterreich-hinaus-erfolgreich/
7. Marvel running account for Iron Man: https://www.facebook.com/ironman/
8. The New Yorker on Fake-News via Facebook: https://www.newyorker.com/podcast/political-scene/how-facebook-continues-to-spread-fake-news
9. The Guardian on Click Farms: https://www.theguardian.com/technology/2013/aug/02/click-farms-appearance-online-popularity
10. Facebook Connect: https://developers.facebook.com/docs/facebook-login/

11. Twitter verified account: https://help.twitter.com/en/managing-your-account/about-twitter-verified-accounts
12. Twitter verifies the right-wing extremist Kessler: https://www.vox.com/2017/11/9/16629796/twitter-halts-verification-white-supremacist-jason-kessler-//policy-blue-check-mark
13. MIT Touchstone: https://ist.mit.edu/touchstone
14. XFINITY on Campus: https://www.xfinityoncampus.com/picker?tenant=x2&continue=http%%2F%2Fwww.xfinity.com%2Fstream%2Faccess-token.html
15. Crptomathic.com about eID Verification Process and Client Onboarding for Banks and Financial Institutions compliant to eIDAS, KYC and AML: urlhttps://www.cryptomathic.com/news-events/blog/eidas-compliant-eid-verification-process-and-client-onboarding-for-banks-and-financial-institutions
16. Intelius – Selfcheck of which public data is available about a person: https://www.intelius.com/
17. Austrian Erste Bank und Sparkasse as identity provider: https://www.sparkasse.at/sgruppe/privatkunden/digitales-banking/apps/s-id-app
18. Mc Kinsey on digital ID as "dual use system" risks and opportunities: https://www.mckinsey.com/industries/financial-services/our-insights/banking-matters/digital-id-the-opportunities-and-the-risks
19. Sovrin Alliance – Self Sovereign Identity Foundation: https://sovrin.org/
20. Phil Windley on the shift from Multi Source Identity (MSI) to Self Sovereign Identity (SSI): https://www.windley.com/archives/2018/09/multi-source_and_self-sovereign_identity.shtml
21. Christopher Allen on Self Sovereign Identity: http://www.lifewithalacrity.com/2016/04/the-path-to-self-soverereign-identity.html
22. History of Österreichische Staatsdruckerei: https://www.youniqx.com/en/mia-my-identity-app/
23. MIA (My Identity App): https://www.youniqx.com/en/mia-my-identity-app/
24. Article about myIDSafe on trend.at: https://www.trend.at/branchen/digital/myidsafe-oesterreichische-log-allianz-start-10715365-
25. EOS hack on cointelegraph: https://cointelegraph.com/news/hacker-spends-1k-to-win-over-110k-in-eos-betting-game-using-rex
26. DAO Hack Ethereum explained: https://medium.com/@ogucluturk/the-dao-hack-explained-unfortunate-take-off-of-smart-contracts-2bd8c8//db3562
27. MIT Technology Review on Blockchain hacks: https://www.technologyreview.com/s/612974/once-hailed-as-unhackable-blockchains-are-now-getting-hac//ked/
28. Blockcerts, the Open Standard for Blockchain credentials: https://www.blockcerts.org/
29. nxt Whitepaper on nxtwiki: https://nxtwiki.org/wiki/Whitepaper:Nxt
30. Ethereum Whitepaper on Github: https://github.com/ethereum/wiki/wiki/White-Paper
31. Deloitte: Are token assets the securities of tomorrow? (2019) https://www2.deloitte.com/content/dam/Deloitte/lu/Documents/technology/lu-token-assets-securities-tomorrow.pdf
32. Thoughts on tokens by Balaji S. Srinivasan: https://news.earn.com/thoughts-on-tokens-436109aabcbe
33. Understanding the difference between coins, utility tokens and tokenized securities by Michael Benoliel: https://medium.com/startup-grind/understanding-the-difference-between-coins-utility-tokens-and-tokenize//d-securities-a6522655fb91
34. Definition by Merriam-Webster of Utility-Tokens: https://www.merriam-webster.com/dictionary/utility%20token
35. PWC Malta on the new regulatory framework in Malta: https://www.pwc.com/mt/en/publications/technology/pwc-malta-blockchain-alert.html
36. Jahresforum Corporate Learning - IMH Conference: https://www.imh.at/veranstaltungen/seminar/e-learning-als-bildungsinnovation/
37. Alexander Pfeiffer Keynote at eEducation Praxistage: Blockchain und AI im Bildungswesen: Wie Blockchain in Kombination mit digitalen Signaturen und Artifizieller Intelligenz (Agents) auch vor dem Bildungswesen nicht halt macht. Ein Einblick in eine mogliche Zukunft.: https://eeducation.at/fileadmin/user_upload/Praxistage_2019_Programm_Detail_Dienstag_final.pdf
38. FAQ on Ardor Wallets: https://ardordocs.jelurida.com/Faq

39. Wang, S., Wan, S., Zhang, D., Li, D., Zhang, C.: Towards smart factory for industry 4.0: a self-organized multi-agent system with big data based feedback and coordination. Comput. Netw. **101**, 158–168 (2015)
40. Oztemel, E.: Intelligent manufacturing systems, smart factories and industry 4.0: a general overview. In: Kumar, K., Zindani, D., Davim, J.P. (eds.) Digital Manufacturing and Assembly Systems in Industry 4.0. CRC Press, Taylor & Francis Group, Boca Raton (2020)
41. Esmaeilian, B., Behdad, S., Wang, B.: The evolution and future of manufacturing: a review. J. Manuf. Syst. **39**, 79–100 (2016)
42. Monostori, L., Váncza, J., Kumara, S.R.T.: Agent-based systems for manufacturing. Ann. CIRP **55**, 2 (2006)
43. Wooldridge, M., Jennings, N.R.: Intelligent agents: theory and practice. Knowl. Eng. Rev. **10**(2) (1995)
44. Huhns, M.N., Stephens, L.M.: Multiagent Systems and Societies of Agents in Gerhard Weiss, Multiagent Systems: A Modern Approach to Distributed Artificial Intelligence. MIT Press, Cambridge (2000)
45. Wooldridge, M.: Intelligent Agents in Gerhard Weiss, Multiagent Systems: A Modern Approach to Distributed Artificial Intelligence. MIT Press, Cambridge (2000)
46. Greene, K.K., Kelsey, J., Frankli, J.M.: Measuring the usability and security of permuted passwords on mobile platforms. Technical report NISTIR 8040, National Institute of Standards and Technology, Information Access Division, Information Technology Laboratory, 100 Bureau Drive (Mail Stop 8940) Gaithersburg, MD 20899-8940 (2016)
47. Pidel, C., Neuhaus, S.: Breaking: password entry is fine. In: Moallem, A. (ed.) HCI for Cybersecurity, Privacy and Trust. First International Conference, HCI-CPT 2019 Held as Part of the 21st HCI International Conference, HCII 2019 Orlando, FL, USA, 26–31 July, 2019, Proceedings, Springer Nature, Switzerland (2019)
48. Jósang, A.: Identity management and trusted interaction in Internet and mobile computing. IET Inf. Secur. **8**, 67–79 (2014). https://doi.org/10.1049/iet-ifs.2012.0133
49. Burden, D., Savin-Baden, M.: Virtual Humans, Today and Tomorrow, Part of Chapman & Hall/CRC. Artificial Intelligence and Robotics Series, CRC Press, Taylor & Francis Group, Boca Raton (2019)
50. Lamport, L., Shostak, R., Marshal, P.: Reaching agreement in the presence of faults. J. Ass. Comput. Mach. **27** (1980)
51. National Center for Biotechnology Information, http://www.ncbi.nlm.nih.gov

Index

A. Dingli et al. (eds.), *Artificial Intelligence in Industry 4.0*,
Studies in Computational Intelligence 928,
https://doi.org/10.1007/978-3-030-61045-6

The manufacturer's authorised representative in the EU is Springer Nature Customer Service Centre GmbH, Europaplatz 3, 69115 Heidelberg, Germany. If you have any concerns regarding our products, please contact ProductSafety@springernature.com

Printed and bound by CPI Group (UK) Ltd, Croydon, CR0 4YY

28/09/2026

02224090-0010